Software Design

Chapman & Hall/CRC Innovations in Software Engineering and Software Development Series

Richard LeBlanc

Seattle University, Washington, USA

This series covers all aspects of software engineering and software development. Books in the series include innovative reference books, research monographs, and textbooks at the undergraduate and graduate level. The series covers traditional subject matter, cutting-edge research, and current industry practice, such as agile software development methods and service-oriented architectures. We also welcome proposals for books that capture the latest results on the domains and conditions in which practices are most effective.

Introduction to Combinatorial Testing
D. Richard Kuhn, Raghu N. Kacker, Yu Lei

Software Test Attacks to Break Mobile and Embedded Devices
Jon Duncan Hagar

Software Essentials Design and Construction
Adair Dingle

Software Metrics A Rigorous and Practical Approach, Third Edition
Norman Fenton, James Bieman

Computer Games and Software Engineering
Kendra M. L. Cooper, Walt Scacchi

Evidence-Based Software Engineering and Systematic Reviews
Barbara Ann Kitchenham, David Budgen, Pearl Brereton

Engineering Modeling Languages Turning Domain Knowledge into Tools
Benoit Combemale, Robert France, Jean-Marc Jézéquel, Bernhard Rumpe, James Steel, Didier Vojtisek

Speed, Data, and Ecosystems Excelling in a Software-Driven World
Jan Bosch

Introduction to Software Engineering, Second Edition
Ronald J. Leach

Software Design, Third Edition
David Budgen

For more information about this series, please visit: https://www.routledge.com/Chapman--HallCRC-Innovations-in-Software-Engineering-and-Software-Development/book-series/CHCRCINNSOFEN

Software Design

Creating Solutions for Ill-Structured Problems

Third Edition

David Budgen

CRC Press is an imprint of the
Taylor & Francis Group, an **informa** business
A CHAPMAN & HALL BOOK

MATLAB® is a trademark of The MathWorks, Inc. and is used with permission. The MathWorks does not warrant the accuracy of the text or exercises in this book. This book's use or discussion of MATLAB® software or related products does not constitute endorsement or sponsorship by The MathWorks of a particular pedagogical approach or particular use of the MATLAB® software.

Third edition published 2021
by CRC Press
6000 Broken Sound Parkway NW, Suite 300, Boca Raton, FL 33487-2742

and by CRC Press
2 Park Square, Milton Park, Abingdon, Oxon, OX14 4RN

First edition published by Addison-Wesley 1993
Second edition published by Addison-Wesley 2003

CRC Press is an imprint of Taylor & Francis Group, LLC

Library of Congress Cataloging-in-Publication Data

Names: Budgen, David, author.
Title: Software design : creating solutions for ill-structured problems / David Budgen.
Description: Third edition. | Boca Raton : CRC Press, 2021. | Series: Chapman & Hall/CRC innovations in software engineering | Includes bibliographical references and index.
Identifiers: LCCN 2020037195 (print) | LCCN 2020037196 (ebook) | ISBN 9781138196612 (paperback) | ISBN 9780367676568 (hardcover) | ISBN 9781315300078 (ebook)
Subjects: LCSH: Computer software--Development.
Classification: LCC QA76.76.D47 B83 2021 (print) | LCC QA76.76.D47 (ebook) | DDC 005.1--dc23
LC record available at https://lccn.loc.gov/2020037195
LC ebook record available at https://lccn.loc.gov/2020037196

ISBN: 9780367676568 (hbk)
ISBN: 9781138196612 (pbk)
ISBN: 9781315300078 (ebk)

Typeset in Computer Modern font
by KnowledgeWorks Global Ltd.

To the late Jim Tomayko.

Historian, software engineer, teacher, sportsman, coach, aviator, author, family man, and an inspiring friend who had a huge heart for others, whatever their role in life.

Contents

Preface to the Third Edition

"*Science is built up of facts as a house is built of stones, but an accumulation of facts is no more a science than a heap of stones is a house.*" Jules Henri Poincaré (1854-1912)

When using this quotation to open the second edition of this book I observed that it is sometimes hard not to feel that our knowledge about how to design software might sometimes feel akin to having a heap of stones. We possess a collection of observations, techniques and experiences, but finding ways of putting them together to provide an organised corpus of knowledge about how to design software is something of a challenge. The first and second editions of this book have sought to gather, classify, categorise and interpret the available knowledge with the aim of providing some sort of structure that will help the reader to understand and use it—and this one aims to do the same.

The ten years that elapsed between the first and second editions of this book saw the emergence of many new and sometimes radically different ideas about how we might go about designing software. The software designer's toolbox acquired concepts such as *architecture* and new forms of software technology such as software services; there was a move away from 'waterfall' thinking with *agile methods* as well as with new forms of reuse through *design patterns*; and there was a greater 'standardisation' of modelling forms with the *Unified Modeling Language* (UML).

In the (rather longer) period between the second edition and this one, although new ideas have continued to emerge, mostly they have been less radical in their scope. Perhaps this has been partly because software developers have also had to cope with significant changes in the *context* within which software design takes place. It is now increasingly the case that applications may well be developed by globally dispersed teams; make use of open source components; be in the form of *product lines*; need to operate within a global network of systems rather than in a purely local environment; and of course, increasingly need to cope with the possibility of malicious attacks. In addition, the growing availability of knowledge that is based upon empirical evidence is slowly giving us a better understanding of what works, when and why.

So, in order to keep the focus upon ideas about *design*, rather than being swamped with detail about software, this edition has been organised very differently to the previous two. In particular, the presentation of material about design has placed a much greater emphasis upon the role of design as related to the concept and characteristics of *ill-structured problems* (ISPs). Software development presents a particularly challenging example of an ISP—and design forms an important problem-solving tool for addressing ISPs (indeed, the main one). Using the idea of the ISP as a framework for introducing many of the ideas involved may therefore help with understanding the roles of many of the topics addressed in the book.

A rather different approach to codifying and presenting knowledge about *software* design itself has also been adopted for this third edition. This knowledge is organised around the way that the word *design* can be used as both:

- a *noun*, as when we refer to 'a design', meaning our description of how a software system should be structured and the ways that its elements should behave and interact; and also as
- a *verb*, referring to the activity of creating a design, as in 'to design' something.

Together, these ideas underpin the way that this book has been organised. Starting with a set of chapters that use the concept of an ISP to help examine design issues in the wider sense, we then have a set of chapters that examine different aspects of the ways that we can describe a software application and its properties. This is then followed by a set of chapters that look at the different activities that can provide the creative processes needed to produce these descriptions. Using such a structure embodies a sound pedagogical principle related to how we learn—we first learn that words have some sort of meaning, move on to read (the words that others have created) and eventually learn to write (and so create our own words). We can learn about designing software in the same way, gaining an understanding of why we need to design software, what a software design is, and what qualities to look for in it, before we can go on to think about how we might set about producing a design.

As well as adopting a new structure for this edition, there are many other changes that reflect our increasing knowledge about, and experiences with, software design (both as noun and verb). Some older material has been omitted, the terminology has been revised in places, and there is a greater emphasis upon the context within which the development of a design takes place. The bibliography has been updated and extended, with particular emphasis upon including knowledge drawn from empirical studies where this is available.

One other change is that the material of the book has been presented in a less 'formal' manner. This is reflected in a number of ways. One of these is the greater use of informal illustrations and figures. When we provide students of design with neatly produced diagrams, as we usually do in books, papers and lectures, we easily do them a disservice. Designs are not produced via drawing tools, they are far more likely to be *sketched* on a whiteboard or a sheet of paper. And all too often, such sketches will not use a particularly formal syntax or semantics either. So in this edition, there is a move away from neatly formatted 'box and line' diagrams to greater use of hand-drawn sketches—and with it my apologies for the accompanying handwriting. This move to greater informality has also been extended to the illustrations intended to provide 'memory aid' via visualisations of various concepts—within the capacity of my draughtsmanship at least!

What has *not* changed is what has usually been considered as the unique strength of this book, in that it seeks to describe the available knowledge about software design in a balanced and advocacy-free manner. In a field such as computing, where enthusiastic advocacy may drown out evidence and evaluation, this role continues to be an important one. This is not to imply that the aim is to create a "dry and dusty tome". The act of designing something is a creative and exciting one, whatever the domain, and is one that readily provokes enthusiasm among practitioners and students—and can often lead to some frustration too, we might admit. This is every bit as true for designing software as it is for designing cars, aircraft, bridges, toasters and televisions—the aim as always is (or should be) to create something new and better for its users.

One other aspect of the book that has not changed is the issue of using material taken from the web. The ephemeral nature of these sources, combined with the difficulty of finding out how up-to-date they are, as well as the issues of provenance for the material they contain, means that I have been rather reluctant to cite other than a small number of those that I have consulted. Where possible, I have tried to cite only those sites that I consider to be relatively stable, and that can also be regarded as being authoritative in a technical sense.

Acknowledgements

My interest in software design now dates back over some forty years (oh dear!), so giving due acknowledgement to all of the people who have influenced my understanding and my thinking would be a very long list indeed. However, I should particularly acknowledge here the contribution made by the late Norm Gibbs when he was Director of the Education Program at the Software Engineering Institute of Carnegie Mellon University. Norm encouraged me to persist with my early attempts to categorise and classify knowledge about software design, starting with my work on developing a prototype for the SEI's *curriculum modules*, which eventually led to the first edition of this book. I would also like to thank Marian Petre and André van der Hoek for making me so much more aware of the importance of sketching and informality, for both the purpose of design and also for presenting ideas.

I would also like to repeat my thanks to all those who helped clarify my thinking for the first two editions: firstly the students who have had to put up with my use of my teaching role to explore new ways of explaining this material; secondly, my family, who have put up with 'yet another book' (and yet another set of promises about 'nearly done'), mentioning particularly my daughter Jane who went through my sketches to check how well they conveyed the issues (certainly not to check their artistic merit!).

And as with the previous two editions, any errors of fact and any omissions are entirely my own (and unintended)!

David Budgen

The City Car Club

Before we engage in thinking about designing things, this is a good point at which to introduce the *City Car Club* (CCC). The CCC provides a running example that will be used through most of the following chapters to help illustrate various issues. While it is by no means the only example that will be used in these, it has the benefit that once described (here) it can be easily used without the need for lengthy explanations.

Very briefly, the City Car Club is a car-sharing system that will be operating within a city. We haven't said which one, so you can assume your favourite city when picturing it in your mind. The team behind the *CCC* will need extensive software support to make the venture work successfully, and an initial specification of what is needed is given below.

The Specification

The *City Car Club* (CCC) is an organisation that is planning to set up a car-sharing system that will eventually operate in different cities, and requires a software system to manage this. They will be trying out this idea for the first time in ⟨*your favourite city*⟩. CCC will provide a number of small 'city-friendly' cars that can be used by members of the club, who pay a monthly membership fee. When using a car, the pick-up and drop-off points must lie within a specified central zone of the city, although a car may go outside this area while it is in use. There will be some designated locations where cars can be left and collected, indicated by specially coloured hatching on the parking bays, and with these being protected by a lockable 'fold-down' post where it is necessary to stop others from using them. Cars may also be left in any other suitable parking area, including a number of supermarket car parks (approval has been obtained for this), but if a car is left in any other non-designated places, the club member who is leaving the car will be personally responsible for any parking fees or fines that might be incurred. A member will also be charged an additional fee if a car is left outside of the designated central zone.

The position and status of the cars is tracked using a GPS system built into each car. At any time each car will have a status, describing its availability, which can have one of the following values:

- *Unavailable* – the car is currently in use by a member of the club, or has been identified by the support team as needing attention in some way.
- *Available* – the car is available for use and can be booked by a club member.
- *Reserved* – a member has requested use of the car, but has not yet taken charge of it.

CCC's central monitoring system will continuously record and track the position and status of each car.

To use a car (a 'session'), a member uses their club id to access the central booking system, identifies a car that is available as well as conveniently positioned for their use, and makes a reservation, with a unique confirmation code being sent to them as an SMS text or in an e-mail. The reservation can be made using a laptop application or a phone app (these are provided when a member registers with CCC). A reservation automatically expires if a car has not been collected within 20 minutes of making the reservation, but it can be extended for one additional period of 20 minutes (the system prompts the member at the end of the first 20-minute period and gives them the opportunity to extend it). On reaching the car they have reserved, the member uses the unique code to open the car and to operate the ignition system, as well as to unlock the security post if one is in use. This can all be done by using either a small keypad that is provided to members by CCC, or by means of a Bluetooth link to the app on their phone or tablet. Entering the code starts the session, and from that point on the member is responsible for the car until they finish the session by locking the car and notifying the central booking system.

Billing for a session is based upon a formula that takes account of distance covered and the length of time the car is being used. Members can also add fuel to a car at a number of authorised agencies (including the supermarkets providing car parking) using a payment card that charges the fuel to CCC, for which they get credit. There is a 'fine' for leaving a car with an inadequate reserve of fuel in the tank (the fuel level is transmitted to the central monitoring system at the start and end of a session). As well as identifying where nearby available cars are located, the phone app provided by CCC can be used to locate nearby filling stations.

The support team of CCC can perform various 'mobile' tasks of maintenance, including fixing any problems with cars and retrieving any that have been left outside of the central area. While out 'on the road', they will use a similar set of apps to those provided for ordinary members, but with additional privileges (for example, allowing them to declare a car to be unavailable when they begin work on it, and available once they have finished).

In the longer term, once the system has been tested out in this first location, CCC plan to make their cars available in other cities. So the design of their software needs to ensure that this is adaptable for use in other places, in addition to the usual requirements for security, robustness and efficiency.

Highlighting the examples

To help the reader identify where the CCC has been used to illustrate an issue, the relevant text and any associated diagrams will normally be placed inside a separate shaded box as in the illustration below.

Using the CCC in an example

When we are talking about how we might use a particular concept with the CCC then it will appear in a box with grey background and rounded corners like this.

And when we are using a diagram or a sketch to illustrate an issue, then it will be included as part of the example, like the one below.

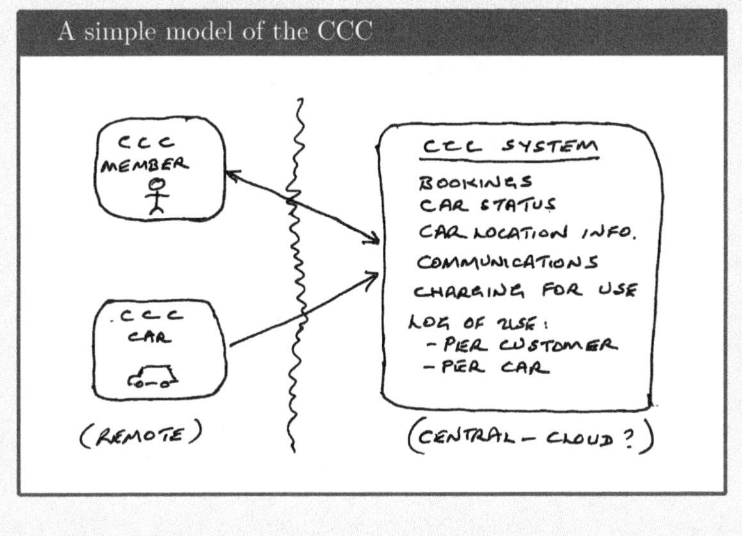

Of course, where we simply make reference to the CCC, rather than discussing how it might form an example of something, we will just do so in the normal text of a chapter.

Sketches versus diagrams

We will say more about what we mean by *sketching* later on, but it is worth pointing out here you will find that many illustrations and figures in this book have been drawn and lettered by hand (admittedly, not always very elegantly). This is because sketching is what designers actually *do* (Petre & van der Hoek 2016), and producing neat diagrams, of the form usually encountered in textbooks, often comes later, if at all.

So, when syntax and clarity are important, diagrams will usually be in 'textbook style', but when the issue is more one of thinking like a designer, then expect to get a sketch!

Part I

Addressing Ill-Structured Problems

Chapter 1

What Is Designing About?

The purpose of this book is to look in detail at how the activities involved in *designing* can be applied to the context of *software development.* Both are challenging topics in their own right, so the chapters that follow draw upon a host of ideas that come from quite an eclectic variety of sources. This chapter is mainly concerned with explaining something of the essential characteristics of *design thinking*, and the various ways in which this is employed during software development. However, since it is largely written from the perspective of "what should a software developer need to know about software design?", we also provide a brief review of the relevant characteristics of software. After all, it is these characteristics that make the task of designing software to be such a unique challenge.

1.1 When is design needed?

In our everyday lives we are surrounded by things that have been designed. The chairs we sit upon, the cars we drive, the phones we use to communicate with others, the houses/flats that we live in. These are all examples of *artifacts*, things that have been created through some form of human activity. All of these have also been created to meet some form of human need and hence can be considered as providing a solution to some form of 'problem'.

In the (relatively) early days of computing, Herb Simon (1973) recognised that computers could be used to help address two distinct types of problem.

- *Well-Structured Problems* or WSPs. These are distinguished by having a 'right' solution, and some clear tests that help demonstrate that this has been found. For example, to find the roots of the quadratic equation $x^2 - 5x + 6 = 0$, we can either use inspection or the standard formula to determine that their values are $x = 2$ and $x = 3$. Putting either value back into the original expression provides a simple and complete test of correctness for that solution. Applications such as financial modelling and database transactions provide examples of using computers to address WSPs.

- *Ill-Structured Problems* or ISPs. For such problems there will usually be no explicit formulation, and many possible solutions, with these being characterised by being good or bad rather than right or wrong. For ISPs there will be no definitive test to determine the correctness of a result, and so no way of knowing that the 'best' available solution has been achieved (usually termed 'no stopping rule'). Computer applications such as face or voice recognition, that have to cope with incomplete information, can be considered as addressing ISPs.

When we look at problem-solving more broadly, not just through the use of computers, these distinctions are every bit as useful. Many problems encountered in science, particularly in physics and maths, are well-structured problems that can be solved through mathematical analysis. However, the problems addressed by engineers, including software engineers, are nearly always ill-structured problems. The need for a bridge can be met in many ways, and there are many ways of writing even a simple computer program. So addressing an ISP involves employing some form of *design* process which enables us to formulate a possible solution and then evaluate how well it might meet a particular need.

The CCC system as an ISP

Let's consider a design scenario that might occur when the software design team start to plan how they will develop the software for CCC.

1. The head designer suggests representing each car as an object, with its current status being stored internally, and a set of methods that allow customers and other users (such as the maintenance team and the accounting software) to access that status.
2. The next question is how to organise all of these objects. So a member of the team suggests using an array of objects. One consequence is that every car now has to have an index number so that the relevant object can be accessed.
3. And then there's the question of how to model the status of a car internally (is it available, unavailable or reserved?).

None of these decisions exists in isolation, and each one has implications for the way that the design will evolve.

Figure 1.1 provides a simple schematic illustration of the different ways in which we need to approach the task of 'solving' ISPs and WSPs. A really important distinction is the way that postulating a possible solution to one aspect of an ISP may influence our understanding of that aspect as well as interacting with the ideas we may have for solving its other aspects (indicated by the dashed-line backward pointing arrows).

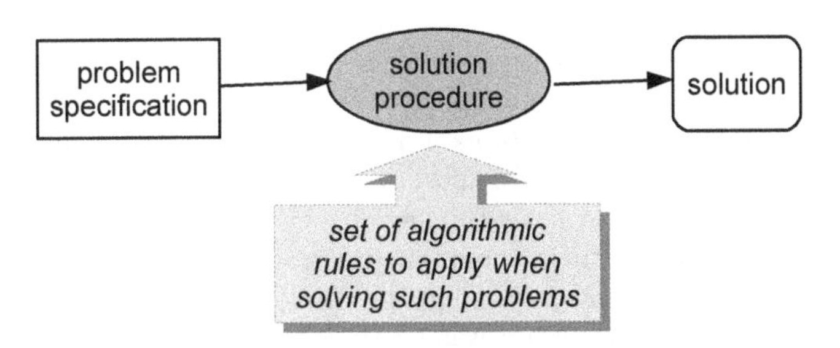

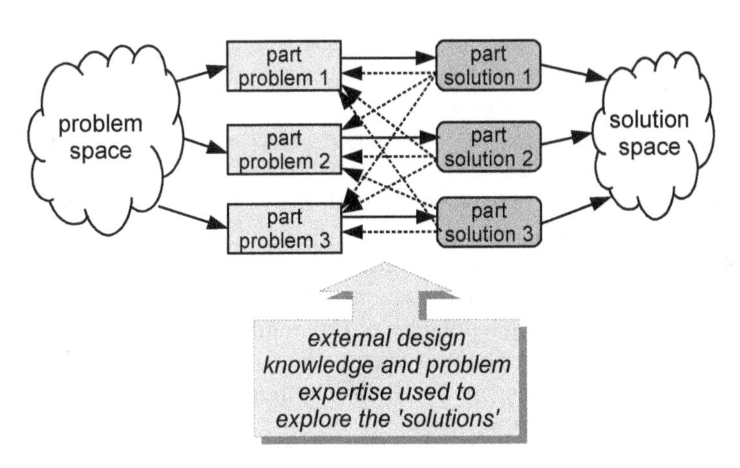

FIGURE 1.1: The solution processes for WSPs and ISPs

So this is really a book about how we tackle ISPs within the domain of software development. We look a bit more closely at the nature of ISPs in the next chapter, and in the rest of this chapter we briefly consider some of the other factors that are likely to be involved in designing software.

1.2 A bit about software

Before engaging with design concepts and 'design thinking', we need to give a (brief) mention to the other word that appears in the title of this book, as it forms an important influence upon the sections and chapters that follow.

In the 'pre-internet' era (and yes, there was such a time), software came in two main flavours. *System software*, supplied with the computer, provided an operating environment and *applications* written by users were executed in the context that this provided. So, for most purposes, anyone setting out to design an application could do so within a fairly well-bounded context that was largely determined by their choice of programming language and the resources provided through the particular operating system. This didn't necessarily mean that the task of design was easier than it is today though, since the designer might well have to design around a lack of resources rather than having to choose between different ones.

From the early 1990s, the internet era began to extend this context. And at the same time, the emergence of new ideas about how to organise and structure software, such as classes, objects and software services required that developers find new ways of creating and organising applications. In particular, when designing an application, it might be necessary to envisage not just how it would interact with other applications on its 'host', but also with applications running on remote computers, where these might or might not need to be developed as part of the design task.

And of course, technology has continued to evolve. Rather than being confined to a local computer, the different parts of an application might now be found hosted in the cloud, tablets, phones or even domestic appliances (the 'Internet of Things'). Systems might well need to be dynamically configured during execution, and the designer may well only be responsible for part of a much more extensive system or application.

And even in the pre-internet era it was recognised that the effort invested in creating software applications meant that an application might have quite a long lifetime in use. It was also realised that an application might need to *evolve* during that lifetime in various ways, so as to meet new needs or changes in its environment (Bennett & Rajlich 2000). (What is now often termed *evolution* was then more likely to be described as *maintenance*, with *adaptive maintenance* describing those changes made to address changes in an application's environment.) So the need to anticipate future developments has always provided an influence (or constraint) upon a designer's ideas.

In more recent times, the associated concept of *technical debt* is one that has introduced a rather different perspective upon design thinking (Cunningham 1992, Allman 2012). Technical debt refers to the future consequences that may arise as a consequence of the current decisions that we make about software development. In particular, it has been observed that, unlike other forms of debt, the repayment of technical debt may well need to be undertaken by those other than the ones who incurred it in the first place. The team that develops a software product may well have no responsibility for its later evolution—so any problems stemming from compromises in the design may well eventually need to be addressed by others.

In many ways, the concept of technical debt highlights a key difference between the 'student experience' and the environment within which the software developer operates. Once a particular student exercise has been completed (and assessed), it is discarded and the student moves on to new topics. So students rarely have to think about other than the very immediate consequences of any software design decisions they make. As such, their work incurs no 'debts'. Unfortunately though, when thinking about software design in almost any other context, the opposite is true!

The CCC and technical debt

Let's consider some possible scenarios for how the business of CCC might evolve and some possible effects that these may have upon the design decisions outlined earlier.

- If the scheme proves highly successful, CCC may need to expand their fleet of cars. *Consequence*: the early decision to use an array of objects might prove inconvenient, although this does depend upon how it was implemented.
- Some customers may need larger cars (even though this is a 'city' club!). *Consequence*: while this can be addressed by using inheritance to create new subclasses of the 'car' class, doing so may create other problems, such as how to meet the need for billing for use according to car size.
- CCC may conclude that they need a more elaborate set of status values, for example: "available but outside of the central zone". *Consequence*: this may complicate or extend the set of methods for each object, depending on how these were specified.

None of these is insurmountable, but it emphasises the point that design decisions, especially those made at early stages, may have consequences for later.

So we might reasonably ask if our ideas about how to (successfully) design software have been able to keep up with these changes and concepts? And of course, the answer is both 'yes' and 'no'. Concepts such as *architecture* and *patterns* have emerged to help us conceptualise our ideas about software design, but equally, because design ideas tend to draw heavily upon experience, our ability to make good use of these concepts always lags behind. Indeed, it might be argued that the emergence of the concept of *agile* development implicitly reflects this process of ongoing change. *Plan-driven* design strategies, as favoured in the pre-internet era, essentially assume a stable operating environment for an application, while agile ones allow designers to adapt their ideas to a changing context, which itself may well be changing (and being clarified) as development proceeds. However, despite this, since design ideas usually stem from experience, technology developments are usually well ahead.

This book therefore aims to provide the reader both with conceptual tools for thinking about how a design might be organised, and also with knowledge about some of the strategies that might be used to produce a design. When thinking about software at this level, the Internet of Things (IoT), the cloud, the mobile phone app, the web application, and the more 'conventional' application have much in common, although the detailed realisations might be quite different. Hence this is not a book about software design for distributed systems, web applications, or phone apps, it is a book that explains different ways of approaching the task of software design. And with that, let's get on with thinking about design...

1.3 What exactly do we mean by 'design'?

One of the complications of the English language is that a word can have multiple uses, and we may often have to work out which meaning is intended by considering the context in which the word appears. The word *design* is just such a word, and in this book we are going to be concerned with exploring a range of key concepts and ideas related to each of its meanings. So it is appropriate to begin by examining just what these meanings are.

Firstly, we can speak of 'a design', using it as a *noun* that describes something that forms a *plan* of how we intend some artifact to be constructed. Indeed, we often speak of a 'design plan' in this context, where the purpose of this is to provide a 'model' of how some form of 'product' should be constructed. This particular meaning is therefore widely associated with the idea of *manufacturing* something, whether crafted by an individual, or assembled in a factory (mass production)[1]. So, we may have a design plan that provides guidance on how to construct a model boat (or a real one, depending on how ambitious we are), or an item of clothing, or a car. The information provided in the plan about how to realise that design might be quite detailed and need to be followed exactly, which is probably a good idea when building a boat or a kit car, or it might be something of a more strategic nature that we

[1]London's 'Great Exhibition' of 1851, which occurred at a relatively early stage of industrial development, was essentially about design, and contained over 100,000 exhibits that were essentially design artifacts—admittedly of widely varying taste and usefulness.

are expected to interpret and adapt, which would be more appropriate when making a garment to fit our own figure.

However, rather than a physical artifact the design plan may describe some form of procedure or process that is to be performed in some way. Again, there are many forms this can take—we can regard a musical score as being a plan for a particular form of process, that is to be followed by the player or players. Again, it may be very detailed, or leave scope for improvisation. Other examples of such procedures might be concerned with form-filling, such as applying on-line for a driving licence or a passport, describing the various things we need to do, and possibly in what order. And of course, we can consider a recipe for making a cake as being a plan that describes a process, but also one that a competent cook will be expected to adapt quite freely.

The two forms of plan can also be combined. As an example, the design for a bicycle may describe both the structure of the bike and also specify the way in which it should be assembled in order to ensure that its physical characteristics are not compromised.

Executing a (musical) design plan

An important characteristic of the use of the word 'design' in this way is that we usually associate it with some form(s) of *description* for the plan. And of course, there are many forms of description, which may employ words, pictures, mathematical forms, musical notation, etc. as well as combinations of these. (Few diagrams are very meaningful without some accompanying text, as anyone who has tried to assemble flat-pack furniture will be only too aware.)

The second meaning, which is closely related to the first one, and probably even more important to us here, is when we use the word design as a *verb*, as in 'to design'. This is concerned with the *process* of designing something—with the goal being to produce a plan. The activity of designing something forms the most creative (and usually challenging) aspect of design as a concept, and indeed, much of this book will be concerned with this second meaning, and with the different processes employed for designing software.

While design plans may vary in terms of how exactly they need to be followed, there is usually a degree of formality associated with their structure that is required to make them understandable. In contrast, design processes tend to be much less constrained, and even following a recommended practice invariably requires some interpretation and adaptation. So we can interpret *design thinking* as meaning thinking about how to design something.

In this book you will find that we are very much concerned with both uses of the word 'design'. While their roles are distinctive, they are not wholly independent either: the form of design plan we want to create may influence the way that we approach the task of designing. Likewise, the way that we approach the act of designing will influence the sort of plan that we produce. And of course, the activities of design and implementation of software lack the clear distinction that usually occurs in industrial manufacturing, since many detailed design decisions may be left for the programmer to resolve during implementation. However, wherever possible we will treat them separately, not least for the sake of our own sanity!

Design as verb creating design as noun

1.4 Three perspectives upon design thinking

The idea of designing is one that pops up in many of the things that we do—not just when developing software—and what is involved will take different forms. Here we briefly look at three quite different perceptions about what the act of designing might involve. They all offer a different perspective upon what *design thinking* might involve, and they all have a place in this book.

If we see the word 'design' used in a newspaper or magazine, then it is quite likely that we will be looking at the fashion pages. (Admittedly, very few newspapers have technology pages.) This highlights the first perspective upon design—which relates to how some form of artifact is *styled*, whether it be clothing, spectacles, mobile phones or electric bikes. Indeed, this is probably what many people think of as being 'design'. It is an important one for software too, for example, when we use ideas from Human Computer Interaction (HCI) and Interaction Design to determine the 'look and feel' of a software application—and hence seek to ensure its acceptability to users.

A second perspective is associated with what we often think of as *planning*. Planning spans a wide range of activities and end-products (think of social planning, game plans, etc.). Again, this meaning is also used in software development, where the activities involved in a development project need to be planned so that effort is used effectively, and where testing needs to be planned so that it is as comprehensive as possible. Although this form is less widely used in software development itself, doing so by creating software from 'components' (including software services) can involve elements of this form.

Both 'design as styling' and 'design as planning' (as we will term them) are largely concerned with how we deploy existing and largely familiar resources. And when new elements appear (such as touch-screens) they are likely to stem from external factors, such as technology developments created by others.

The third perspective is rather different, and we can term it as 'design as adaptation'. This is concerned with how we create *new* ways of doing things, often making something possible that couldn't have been done before, usually by adapting and modifying ideas we have used before on some other task, and sometimes combining these in new and different ways. Many ideas we examine in this book provide models that are intended to be adapted, such as *patterns* and *architectural style*. Often the degree of adaptation is fairly limited—if we have found an approach that works, we are unlikely to want to change it too radically—but there will still be times when our ideas have to be modified quite extensively to meet the needs of a particular problem. Figure 1.2 provides a very simple illustration of this concept. At one extreme, someone designing a new database system might well be able to reuse and adapt the experiences of previously developing a very similar one, and so the designer can largely reuse their design model. At the other extreme, designing a completely new application may involve bringing together a range of previous experiences and well-known design forms and then adapting them to work together to address the need.

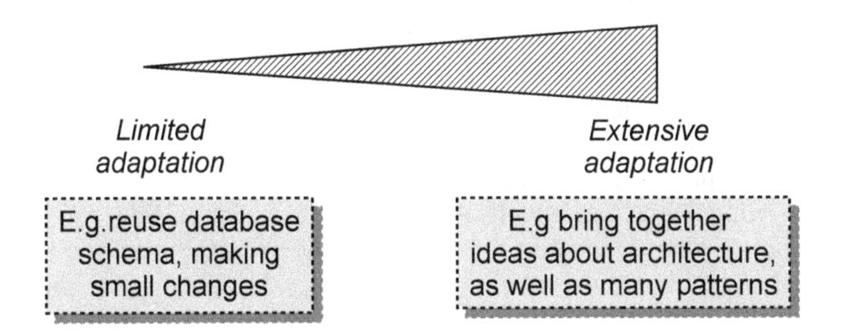

FIGURE 1.2: The spectrum of adaptation involved in designing

All of these three forms embody the concept and characteristics of *design thinking* that we explore in the rest of this book, albeit in different ways. While for software development design activities are often likely to relate mainly to the third perspective, there are going to be elements of all three perspectives in many of our examples. As such, they are not distinct concepts, but more a matter of having different ways of viewing things.

Design perspectives and the CCC

Designing the software for the CCC will mainly be about design as *adaptation*, bringing different forms together to create the models and structures required to meet the needs of CCC. However it will also involve some design as *styling* when considering things like user interfaces that need to be easy to use on a mobile phone. Design as *planning* will mainly be concerned with how the development process is organised—for example whether or not we employ an *agile* approach to development, although there will be some interaction with the first form when considering how the business might develop and expand.

So, what is it that characterises *design thinking*? We explore this more fully in the following chapters, but in essence it involves thinking *creatively* about *how* a task is to be performed. Implicitly, design (in any domain) is concerned with finding ways to do things. That in turn requires an ability to envisage how something might work before we set about trying to construct it. For 'design as styling' that might involve having a vision of how something might appear and/or behave, and then trying to work out how we can bring that effect about. For 'design as planning' we may be concerned with how to use available resources to bring about some result. And for 'design as adaptation' we are likely to be trying to identify a set of mechanisms that will produce the desired effect. The key requirement is one of thinking 'out of the box'—envisaging the end effect that is wanted, and then trying to work backwards to see how that might be achieved.

Design Thinking

And this isn't an easy thing to do. While in principle, anyone can create new things, those things have to meet a range of expectations: they must achieve fitness for purpose, be efficient (in whatever way that matters), have an 'appeal' to users, and many other attributes as well. Somehow, a good designer can draw all these things together while keeping a balance between them—and while we might recognise that they can do so, that doesn't mean that it is an easy task for them, or easy for us to learn to do the same.

So, as we explore design thinking in this book, the aim is to provide a deeper understanding of what makes for good design (largely in a software context, but sometimes more widely too). That doesn't mean that every reader will become a great designer—but at least you can hope to become a competent one. And indeed, most of us will be satisfied to feel confident that we know enough about software design issues to be able to avoid developing software lemons!

Key take-home points about what designing is about

This first chapter has swept across a fairly wide set of issues related to design itself. Here we very briefly recap some key concepts that will be particularly useful when reading the later chapters.

Ill-Structured Problems (ISPs) are characterised by a lack of true/false properties—and are the type of problem that needs to be 'solved' through the use of design rather than analysis. (And software development is definitely an example of an ISP.)

Technical debt is a concept that distinguishes software development that is undertaken 'for real' from software that is written as a learning exercise and then discarded. Designers need to think about the consequences of their decisions in terms of their effect upon the future evolution of an application.

Design can be used as a noun. Employed in this way, 'a design' means a *plan* for how some goal is to be achieved, whether this be physical (how to construct something) or a process (how to perform something).

Design can be used as a verb. This refers to the act of designing (producing a plan).

There are different perspectives upon design. Design thinking applies to a range of activities that can all be considered as being design in some form. Three major perspectives on design that are all relevant to software development are design as *styling*, design as *planning* and design as *adaptation* (experience and models).

Designing is creative. When undertaking design activities, our objective is to create something that does not already exist, which involves predicting an effect upon the future that will only be realised if the design plan is feasible. To do so, we need to be able to envisage how something will be done, and then work backwards to devise ways of bringing this about.

Chapter 2

Doing Design

The previous chapter introduced the idea of *design thinking* and its influence upon software development. In this chapter we look at what it means to apply this to produce a design 'solution'. In particular we examine the influence on this of the *ill-structured* (or *wicked*) nature of design problems and consider how this might affect the ways that we set about 'doing design'.

2.1 Designing as a creative process

A distinctive characteristic of humans is the ability to use tools to create new artifacts, where these are new in the sense that we end up with something that did not exist before. The meaning of 'artifact' is "something made by humans", so distinguishing an artifact from the things found in the natural world. Indeed, the range of artifacts seems almost endless: clothing, books, furniture, buildings, bridges, ships, machinery, computers...

And those are simply the physical forms of artifact. There are also many examples of procedures that humans devise in order to add structure to our lives, whether it involves the process for getting married, organising the way that passengers should board a plane, or negotiating an overdraft with our bank. The tools used to create these are likely to include a variety of conceptual forms (such as workflows and legal frameworks)—but the procedures that result are still artifacts, helping to organise our lives and businesses in an ever more complex world.

The material in the chapters that follow this one is concerned with largely conceptual tools such as design methods, software architecture and software patterns. Software tools for producing more 'physical' aspects of design methods such as diagrams and specifications do exist, but since their purpose is mainly one of *recording* design decisions rather than helping to create new models, they do not really fit into the scope of this book.

Creativity requires more than the availability of conceptual tools of course. What creativity involves is the ability to "think outside the box" and to find new and effective ways to meet a particular need. While this may occasionally involve coming up with a radically different way of doing something, quite often the creative element will be realising that doing things a bit differently might produce something that is more effective, robust, elegant, or some combination of such attributes.

Our perception of how design is involved in producing new ideas will vary, and probably not always be correct. No-one is likely to deny the importance of using rigorous and well-proven engineering design practices when creating motorway bridges, aircraft and buildings, not least because of the safety issues associated with their role and use. Yet good design is important for everyday things too (Norman 2002), even if the effects of poor design lead to nothing more than the odd stubbed toe, or the irritation of having to restart a transaction on our phone or tablet because some information has been 'lost'.

A second, and associated, human characteristic is that of *communication*—although in a design context its importance may arise more from the variety of ways in which it is employed.

Communication with the team

Communication usually plays a vital supporting role to creativity, since developing any new form of artifact almost always requires us to formulate in some way the concept of what this artifact is to be and then to tell others about our ideas. Communication has always been necessary to gain approval and assistance from others, whether it be for the purpose of persuading them to join in with building a megalithic barrow; obtaining a commission to paint a mural for a medieval pope; creating a new social welfare system; constructing a steam locomotive; or creating an on-line banking system. And the act of communication can take many forms, spoken or written words, scale models, mathematical formulae, sketches, simple prototypes etc. And obviously it is closely related to the issue of how we represent our ideas.

Communication also plays another, rather different, role in creative activities. This is to act as the vehicle by which experience and expertise about how to do something (in our case, design software) can be conveyed from an 'expert' to a 'novice', as well as be shared among a community of experts. Indeed, the ability to *reuse* ideas is a major element in the development of design expertise, and it is the systematic codification and organisation of knowledge for the purpose of its reuse that forms an important characteristic of both *craft* and *engineering* disciplines, although these may differ in the degree of formality involved in the codification of their bodies of knowledge.

Both creativity and communication are core concepts that underpin the ideas presented in this book. Indeed, one of the roles for a book like this is that of communication, seeking to explain to others a range of ideas about design and its realisation in a software context.

2.2 Ill-structured problems

As explained in Chapter 1, software development tasks are examples of ill-structured problems (ISPs). ISPs cannot be 'solved' by straightforward analytical means. Well-structured problems (WSPs) can often be solved by using a reductionist approach, gradually transforming the (possibly quite complex) model of the problem into a model of the solution. A good example of this is mathematic models of three-dimensional motion, where the use of an appropriate system of axes can separate out the components so that they can be solved independently. In contrast, ISPs require some form of design process, where this usually involves iteratively postulating a (increasingly complex) 'solution' model, and then evaluating how well this meets the requirements. We discuss this approach more fully later.

An example of the problems posed by an ISP is demonstrated by the way that *feature interaction* can occur when modifying software systems that need to support many options in their operation (such as the software for a phone). Feature interaction is something that arises when the act of adding a new feature that was probably never even envisaged in the original design model causes some of the existing features to stop working correctly. While it may be possible to use analytical techniques to identify where the interactions occur, a resolution of this may quite possibly require a major redesign of internal data structures and the software used to access these.

Using a slightly different terminology drawn from 'design as planning', design problems can be regarded as being examples of what is termed a *wicked problem*. This term was coined by Rittel & Webber (1984) and arose from their analysis of the nature of social planning problems. A wicked problem demonstrates some interesting properties, and it can be briefly summarised as being a problem for which its form is such that the act of finding a solution for one of its aspects will also have the effect of changing our understanding of the problem[1]. (For the discussion of this section, we can largely interpret the term 'solution' as meaning a design plan.) For the purposes of this section we will continue to use Simon's terminology and refer to such problems as ISPs.

Rittel and Webber identified ten distinguishing features of ISPs/wicked problems. Most of these can be equally well interpreted as applying to software systems (Peters & Tripp 1976). Here we briefly review those characteristics of an ISP that relate most closely to software design, and interpret them for a software related context.

- *There is no definitive formulation of an ISP.* The challenges involved in specifying what software systems should do are well known, and indeed, because both specification and design can involve modelling of some required functionality as we illustrate in Figure 2.1, they can be difficult to separate clearly. Rittel and Webber make the point that the activity of specifying and understanding a problem or need is also prone to be bound up with our ideas about how it might be solved. This of course is one reason why simple life-cycle models in which the activities of specification (by the customer) are performed in advance and form the basis for the task of design (by the developer) are usually unrealistic.

- *ISPs have no stopping rule.* This emphasises the point that there are no criterion that we can use to establish when *the* solution to a problem has been found, such that any further work is unlikely to improve it significantly. When designing software it lacks any characteristic we can measure and use to demonstrate that a particular design model is 'complete'. So, rather inconveniently, we have no reliable way of determining when we should stop working on a design task.

- *Solutions to ISPs are not true or false, but good or bad.* In science, we can often identify whether a 'solution' is correct or not, perhaps mathematically, or experimentally. In contrast, there are very many ways of writing bits of software, or structuring them, which will all produce the intended ('correct') outcomes. They may differ in many ways (length, clarity, speed of execution) while performing the required task successfully. (Anyone who has marked student programming exercises

[1]Those familiar with Sellar & Yeatman's classic work "1066 And All That" (1930) might feel that this idea is aptly illustrated by their description of Gladstone's frustration with trying to guess the answer to the Irish Question, and their explanation that "unfortunately, whenever he was getting warm, the Irish secretly changed the Question".

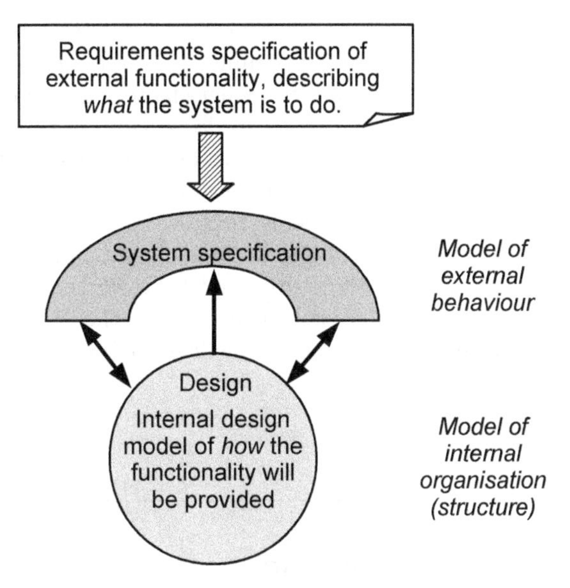

FIGURE 2.1: The interaction between requirements and design models

will be familiar with the wide variation of style and structure commonly encountered—even for the programs that work!)

- *There is no immediate and no ultimate test of a solution to an ISP.* What this effectively observes is that we can't really be sure that what we have produced meets all the identified needs by means of any simple form of test. As we will see, evaluation of a software design is very challenging, and even apparently simple comparisons between design solutions need to be made by using multiple criteria.

- *Every ISP is essentially unique.* In later chapters we discuss ideas about how we can reuse design experience for part-designs through forms such as *design patterns* and *software product lines.* But even with these, the designer needs to adapt and interpret the idea in order to apply them to a specific requirement—which of course is why design activities do not lend themselves to being automated.

Taken together, these ideas are certainly at variance with the reductionist approaches to problem-solving widely employed in mathematics and science. Such an approach seeks to break a problem down into smaller, more manageable (and solvable) elements, with an implicit assumption that there will be a single solution that this process will converge on.

In contrast, ill-structured design problems require the designer to juggle many different aspects of their 'solution' simultaneously. This then requires the development and use of complex models, and making a 'wrong' design choice earlier in the design process may make it harder to resolve later choices. The concept of *orthogonality*, whereby the parameters of a model are independent from each other, doesn't apply to an ISP, where factors may interact in different ways and the best that we can hope to do is to make trade-offs between them. Our choices between options are apt to be driven by the need to find a balance, rather than expecting to find the 'right' outcome.

So, these are the types of problems that we will be addressing in the following chapters. Successful designing does not result simply from following prescribed processes or reusing structures that have worked for others. As we will see, these things might well be useful elements in the design process, or in learning about it, but even then, their use will probably involve some degree of adaptation.

2.3 What does a designer do?

So, what exactly is involved in designing something? A good starting point is to consider the words of a pioneering 'design methodologist', J. Christopher Jones, taken from his classic work, *Design Methods: Seeds of Human Futures* (1970).

> "*The fundamental problem is that designers are obliged to use current information to predict a future state that will not come about unless their predictions are correct. The final outcome of designing has to be assumed before the means of achieving it can be explored; the designers have to work backwards in time from an assumed effect upon the world to the beginning of a chain of events that will bring the effect about.*"

This statement describes an approach to tackling an ISP that is radically different from the reductionist practices of 'scientific method' commonly used with WSPs, and involving subdividing a complex problem into simpler ones. So in its way the statement from Jones is both challenging and unsettling—we are asked to postulate the form of the desired end result in order to try to achieve it, rather than to proceed through a process of deduction and analysis to determine what its form should be.

Science seeks to reduce complexity

One reason for this difference of philosophy is because science is largely concerned with studying things as they *are*, using observation and experiment as the means of testing ideas. A rather simplified view of the scientific process goes as follows: a scientist observes some phenomenon, builds models of how they think this arises, and then seeks to verify and test those models by using them to make predictions, which are then compared with reality through further observation and experiment. (Of course, this will often be iterative.) In contrast, the description of the design process outlined by Jones is one that is more familiar to engineers, and is one that is seeking to create *new* things. The focus is upon devising some mechanism that will have a new effect upon the world, rather than upon analysing something that exists. To achieve this, we need to postulate what the end result will be, and then to try to devise the means of bringing it about. So this is where creativity comes into play!

Implicitly, such an approach contains some degree of uncertainty, mainly because the 'solution space' for an ISP is potentially very large. And this uncertainty occurs partly because we are usually tackling the whole problem, not trying to use a reductionist approach to resolve it as a set of constituent parts. As a result, we may fail to find an effective means of bringing an effect about, or at least, our early attempts might not find the best ways to do this. (Think about some of the early designs for cars, aircraft, television sets etc.) Although of course, an appreciation that these are not the best ways might not be obvious at the time. And while scientific deduction can lead us down blind alleys too, there are usually better tests that can be used to identify and recognise these than are available for assessing the practicality of design solutions.

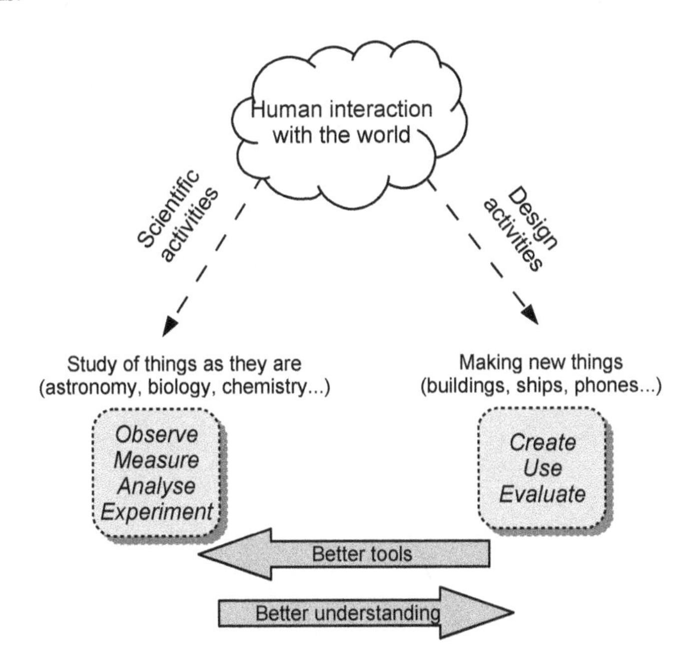

FIGURE 2.2: How science and design interact with the world

Figure 2.2 shows the difference between these two processes in a schematic form. Both are gross simplifications of course, but they focus upon the essential differences, not the detailed nature. And the implication that the role of science is to provide a better understanding of the world, while that of design is to provide better tools for living in it, is also something of a simplification.

If we examine the quotation from Jones a little more closely, and rephrase it a little, we can identify some of the actions that a designer will need to perform when deriving and specifying a 'design solution' (usually termed a *design model*) to meet a given need. And, as there will be many possible solutions, the designer needs to refer to "a solution" where the scientist can usually refer to their goal as "the solution". These actions include:

- postulating how a solution may be organised;
- building a model of the solution (the design model);
- evaluating the model against the original need;
- elaborating the design model to a level where it can be realised.

Of course it is never as simple as this! While there is an implicit ordering for these activities, they are by no means likely to be performed as a simple sequence. It is usually necessary to perform many iterations at each stage as well as between stages, and quite extensive backtracking may be needed when design choices have to be re-evaluated. Indeed, as Jones (1970) recognises, the position may be even worse, since:

> "*If, as is likely, the act of tracing out the intermediate steps exposes unforeseen difficulties or suggests better objectives, the pattern of the original problem may change so drastically that the designers are thrown back to square one.*"

In other words, the act of elaborating the original model may reveal its inadequacies or even total unsuitability, requiring that it be discarded altogether!

Discarding a design idea

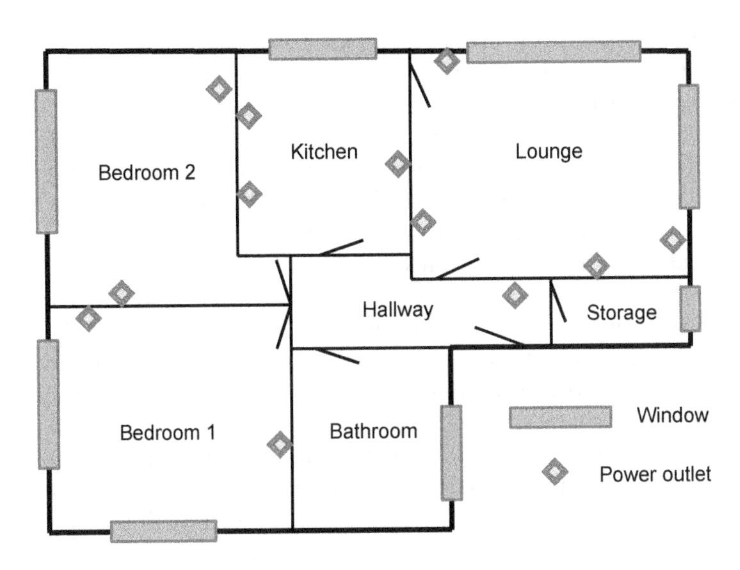

FIGURE 2.3: Plan of a single-storey house

2.4 A simple example: the house move

At this point, it may be useful to look at a simple illustration of the use of design thinking. Moving to a new house or flat is a fairly common experience, and is widely regarded as one of the major traumas in life. However, within this lies a good illustration of the use of design thinking, so for our example, we briefly examine some of the activities that might be involved when planning a move into a new house. Strictly of course, this is 'design as planning' rather than 'design as adaptation'. However, using this perspective provides a rather easier introduction to, and illustration of, the nature of designing as well as of ISPs, while possessing the same key characteristics.

One of the practices often recommended for organising a move is to begin by measuring the dimensions of the various rooms in the house, then to obtain some squared paper and use this to draw a scale plan showing all of the rooms, with the positions of doors, windows etc. (If this sounds a bit low-tech, just wait...) Figure 2.3 shows a simple example of this. The next step is to measure the dimensions of our furniture and then to cut out pieces of cardboard with shapes that represent this on the chosen scale. Together with the scale plan, these form a 'model' of the house and its future contents.

The following bit of the process is the 'design' element, which is also pretty low-tech, and involves trying to identify the 'best' positions for all of our furniture in the new house by placing the cardboard shapes on our model and trying to visualise the outcomes. There are some obvious constraints (we don't usually want bedroom furniture in the kitchen), and there may be issues of colour or style which affect how specific items of furniture might go together, but essentially there is really no systematic way of doing this. Nor is there a clear measure that tells us that we have found the 'best' plan. And of course, this is really a two-dimensional representation of a three-dimensional situation, so we also have to consider the possibility of furniture blocking windows or radiators, or obstructing access to power sockets [2].

The decisions that result from this process can then be regarded as providing a *plan* that can help the removal team on the day. Indeed, not only will it tell them where items are meant to go, it may also affect the way that they pack the items in their van so that unloading is made easier. So, the outcomes from our design activity may well feed into the way that they go about planning their tasks too.

Compared with designing software, this is a relatively simple process. Yet it is sufficient to illustrate some of the ways that design differs from analysis. As noted previously, when solving WSPs, especially when these are mathematical in nature, we try to separate different variables and simplify our task by deriving a solution for each of them separately (reductionism). In contrast, for design we have to manipulate all or most of the elements of a complex model throughout the process. There are also many possible 'solutions'; there is no really useful criterion to tell us that we have found the optimum one, and no way of knowing when to stop (other than exhaustion). And there may be trade-offs too: we may need to accept the odd partially-blocked power socket in order to get the layout that we want in a particular room. Finally, it is still a (design) model—on the day, when we see everything in place, there's a good chance that it will become blindingly obvious that it is possible to improve substantially on what we have devised, or even worse, that there is a major flaw in it.

However, in many ways, this provides a much simpler environment than the one that the software designer faces. The model is closely and physically related to the solution, the set of objects being manipulated is a fixed one, and the properties for each of them are known. Also, the design process only involves us in modelling physical objects rather than processes, unlike designing with software. So, while the example illustrates some key aspects about the design process, we should not forget that things are rapidly going to get more complicated when we come to design software applications.

[2]The reaction of many software engineers is probably to suggest building some sort of software tool to assist with this, that could help by highlighting any constraints that are violated. Useful though this might be, it doesn't do anything to change the fact that we still have to make all of the decisions about where things are to go.

Key take-home points about designing

This second chapter has largely focused on how *design thinking* is applied to designing new things.

Designing is creative. The aim of designing is to create something that does not already exist. This requires the designer to predict a (desired) outcome and explore ways of bringing this about.

Design is used to address ill-structured problems (ISPs). This means that there will be a large number of possible 'solutions' (designs) for a given problem, and that the process of designing has no 'stopping rule' that tells the designer that the current state of their design solution is good enough. As an added complication, there are no simple measures that can be used to assess the feasibility of a design plan.

Designers solve problems differently to scientists. Scientists commonly address problems by following a process of *reductionism*, seeking to resolve a complex problem into separate simpler ones that can be more readily solved analytically. Designing involves exploring all, or most, of the characteristics of the 'design model' at once, rather than being able to consider them separately.

Chapter 3

Managing the Design Process

Given that design (as a verb) is an important means of developing 'solutions' to complex ill-structured problems, an important question is how we manage this process, and what information we need to record as part of that task.

3.1 Cognitive capacity

Back in 1956, George Miller, a psychologist from Harvard University, published a paper titled *The Magical Number Seven, Plus or Minus Two: Some Limits on our Capacity for Processing Information* (Miller 1956). Essentially, this very extensively-cited paper was about the limitations of human cognitive capacity for processing different forms of information. Strictly speaking, Miller's main focus was upon the ability to make judgements related to distinguishing between different stimuli of the same type, such as sound tones, or tastes, although he also discussed the quite distinct limitations upon immediate memory. However his use of the number seven as representing some form of limit on cognitive capacity has subsequently been interpreted (or misinterpreted) very liberally.

The key point for our purposes here is to appreciate that the human brain has a limited capacity for recognising things on a 'one-dimensional' scale. Indeed, the exact number of things that can be distinguished is not too critical as far as we are concerned; what matters is that there is a *limit* on how much information about a design can be readily accommodated in the designer's head, and that the limit is set quite low.

In practical terms, what this means is that managing the design process requires that the designer needs to use some way of handling very many design elements that each have complex properties. It simply isn't possible to hold everything that needs to be considered in one's head while developing or modifying a design model, meaning that the designer needs tools to help them cope with these limitations upon cognitive capacity. (And of course, the tools themselves may involve cognitive limitations. As we will see, diagrammatical notations do seem to be harder to use when they have overly many symbols—a point that the designers of the UML do not seem to have grasped (Moody 2009).)

So the rest of this chapter looks at some of the ways that we can cope with these cognitive limitations, and what it is necessary for a designer to do.

3.2 The power of abstraction

In order to create manageable models of the intended system and envisage how it will behave, we need to find ways to put aside the details that don't matter and to focus upon the factors that do matter.

The concept of *abstraction* is an essential one when formulating design ideas in any branch of engineering. The process of abstraction involves removing or omitting unneeded detail from a description of a model or plan, while still retaining a description of those properties that are relevant. Doing so reduces the cognitive load involved in understanding particular aspects of a design model.

Our earlier example of moving house illustrates this well—the two-dimensional plan provides a model of the house that retains only the information that is important to consider when positioning furniture. It describes the things we need to know for this purpose (where sockets are located, which ways the doors open etc.) but omits such information as whether windows are clear or opaque, what colours the walls have been painted, or where the lights are located etc. Of course, for other tasks we might need a quite different abstraction of the house. If we are planning to modify the power outlets and lights, we would need a very different sort of representation that relates to the different electrical circuits, and that probably omits much of the information considered appropriate for positioning furniture.

Abstraction is both an important concept and also one that novice software designers may well find difficult to employ effectively, as demonstrated in an early study by Adelson & Soloway (1985). Programmers are used to working with what in many ways is a wonderfully pliable medium, but the nature of programming requires that the programmer needs to anticipate all of the different ways that a chunk of software will need to work. It is only too tempting for the novice designer to focus upon the 'comfort zone' of detailed programming structures when thinking about design, since these are likely to be more familiar, whereas as a designer, they need to learn how to think about the properties of a system in an abstract way. Design models need to relate to events, entities, objects or whatever other key elements are involved in thinking about a particular aspect of the design, and to leave to a later phase issues of detail such as the choice of specific data structures, or the form of loop construct needed for a given algorithm etc.

Abstraction therefore plays a key role in this book, corresponding to its central role in designing systems. In particular, when we look at how the characteristics of a design can be represented we will find that we need to think about a number of different abstractions. The effective and appropriate use of abstraction is a key skill that any designer needs to learn and employ. We will be revisiting this in the next chapter and in many of the later chapters.

As a final thought about abstraction, and to provide a tangible illustration, we briefly revisit our first sketch related to the CCC.

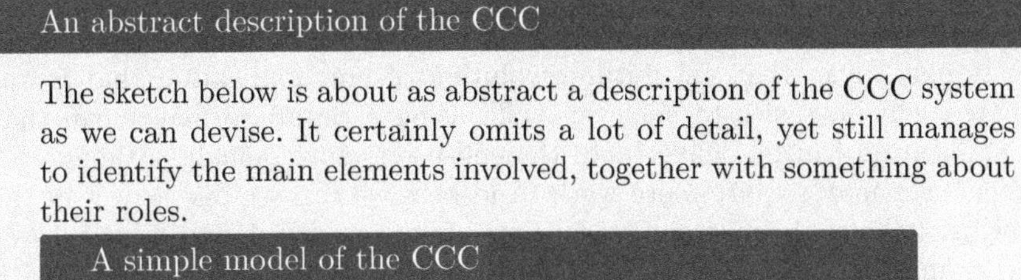

An abstract description of the CCC

The sketch below is about as abstract a description of the CCC system as we can devise. It certainly omits a lot of detail, yet still manages to identify the main elements involved, together with something about their roles.

A simple model of the CCC

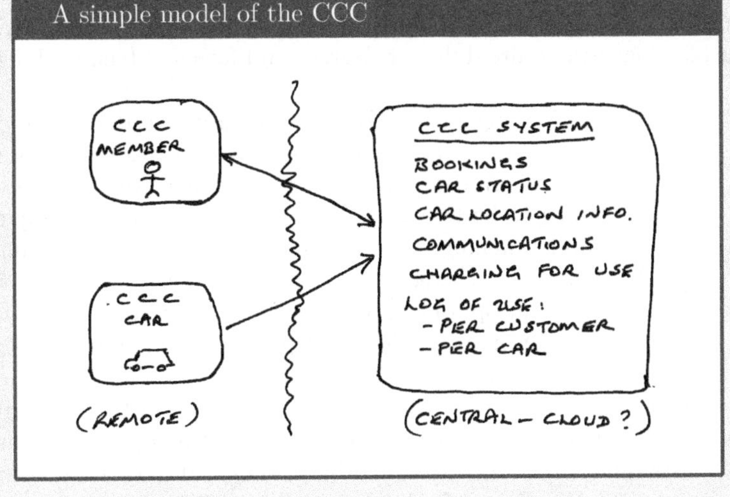

3.3 Modelling and making design choices

When describing what designing is about, we might say that our aim is to produce a 'design solution' that will meet a specific need. In computing, that need will typically be documented in some form of *requirements specification*. This may be little more than a few jotted notes or sketches, through to a very detailed formal document (the specification for the CCC system comes somewhere between those). It is then the task of the designer or design team to formulate a way of meeting that need by creating some form of design model for the proposed solution.

The act of designing something usually involves identifying suitable options for the key features of our 'design solution' and then evaluating the consequences of adopting each of them. For example, one of our first decisions may well be to choose an appropriate *architectural style* for our application. (We will discuss the idea of system architecture and architectural style more fully in later chapters; for the moment we can just consider the architecture of an application as being the types of software module we will employ.) So, for a particular application, we might consider organising it using an object-oriented form, based upon a set of classes and objects, or organising it around a central database. Both approaches will have their merits: using objects may mean that any future evolution of our application that is anticipated will be easier to achieve; while using a database may mean that the application can be delivered to the end-users more quickly, although future changes may require more work. And here we can see one aspect of the problem of choice, in that our reasons for preferring one design solution over another may well be based upon how we rank quite different criteria, with the options also incurring quite different degrees and forms of technical debt.

Making a choice between the different design options may also be based upon criteria that are relatively ill-defined. And as the early choices we have to make are going to be based upon a very abstract and incomplete model, making them will almost always involve making trade-offs between estimates of the likely impact of the different options upon both system functionality as well as such non-functional properties of the model as size, speed and ease of change. There may also be other more context-specific factors that might influence design choices: in that an application may need to conform to the way that an organisation operates; it may need to work with, and communicate with, other systems; and there will nearly always be the need to plan for likely future extensions. This last point, with its implications of different levels of technical debt, reminds us that designers need to think to the future, not just to the present.

Throughout this process though, the designer needs to keep in mind that the ultimate criterion has to be *fitness for purpose.* It is not that the other factors don't matter, they certainly do, but they need to be subordinate to the need to produce a system that does the required job. This goes for most artifacts of course—however much we might like the added "bells and whistles" of a particular model of car or of a specific hi-fi amplifier—pragmatism usually means that our ultimate choice will need to be made on the basis that it can be relied upon to do the job that we need it to do, and that it will do so at a reasonable cost.

Using these criteria to help evaluate possible options can be made easier if the designer produces some form of *design model* that can be used to help manage the cognitive processes involved. The nature of an ISP means that we need to postulate a design solution and then reflect on whether or not it will do the job we want, and how well it will do it. Having some form of abstract design model, possibly realised in the form of diagrams and text, can help with managing this process of reflection, perhaps by using it to evaluate different scenarios of use or change via some form of 'design walk-through'.

And finally, the idea of fitness for purpose isn't some abstract concept, it needs to be given concrete values that can be used to assess the terms above. We need to have a clear picture of what we expect of something that 'should work' and of what we mean by 'as well as possible'. If we are building something that is safety-critical (that is, lives will be at risk if it fails), then we should have some quite definite ideas about these terms. This is less so where the concept of 'works' might be less precise (such as an application that depends upon some form of image recognition), where we might be willing to accept a less than complete level of reliability, at least in early versions.

This is maybe where some reverse thinking may help. Rather than trying to define 'should work', we might try to consider what sort of shortfalls in functionality or performance might still mean that we consider that the system 'works', even if it is not doing so as well as we wish. For the example of image recognition, we might be willing to accept that our application may sometimes be unable to recognise an image, or fail to get the right match, particularly if such a failure can be easily identified by the user. On the other hand, an application where the outputs are occasionally badly in error in a way that cannot easily be recognised by the user is likely to be considered unacceptable, even if this occurs very infrequently.

3.4 Recording design decisions

Something that is often overlooked when a software application is developed is the value that can be derived from noting *why* particular design choices were made. While the rationale for these may have been obvious to the designers of the original system, they may well be far less obvious to those who are later tasked with extending or modifying a system. However, while our design model will usually record *what* choices were made, the *rationale* behind those choices is not so readily included. And although the absence of knowledge about the original rationale can add significantly to technical debt, it often remains overlooked. (In fairness, this problem is not one that is confined to software engineering; it happens in other forms of engineering too.)

Beginning with the original design, the recording of rationale is more likely to be encouraged if the software development project includes some form of *design audit.* Such an audit may consist of a peer review of the designer's ideas, or it may be something rather more formal. Whatever form it takes, an audit may help with recording the reasons for particular design choices. And conducting such an audit requires that we have a design model that is recorded in a form that can be shared with others.

Arguably there is an equally good reason for the rationale for maintenance (design evolution) changes to be recorded. Software applications may have quite long lifetimes and undergo extensive changes over that period, often involving much greater developer time than the original development. Ideally, the maintenance team need to be able to recreate the original models used by the designers so that they can reliably decide how to implement their changes in a manner that preserves the structure of the application (Littman, Pinto, Letovsky & Soloway 1987). In turn, they too should record the reasons for their changes.

Unfortunately, while different software design approaches generally encourage the recording of decisions about how an application should be structured, they are apt to place less emphasis upon recording the rationale for those decisions. It is easy to understand why this is so. A design usually evolves through a set of different stages and forms, and recording the reasons for each step and/or change in thinking may well be seen as a chore by the design team. This issue was recognised many years ago by Parnas & Clements (1986). They observe that, even if the design process was not a rational one (which it usually isn't), the documentation that described it could still make it appear as though it were. In other words, the design documentation should be written so as to describe a consistent and rational process.

Parnas & Clements argue that the principal benefits of doing so are that new members of a design team (or those later charged with extending the design) will then be able to absorb knowledge about the project more easily. Indeed, they observe that even for scientific studies such as mathematical proofs, the form eventually published is rarely the form in which the proof was actually derived, because as understanding grows, various aspects can usually be simplified. In the same way, it is the actual structure of the system design, and the reasons for it, rather than the way that these emerged, that will be of importance to new team members.

An excellent example of the complexity of team development is that of many open source software (OSS) projects. Although various tools exist to help track a project through its many iterations, there still seems to be relatively limited scope (or perhaps motivation) to record design rationale, which in turn can provide a challenge to anyone joining a project (Steinmacher, Silva, Gerosa & Redmiles 2015). And for professional developers, recording decisions and rationale is still only likely to occur where there are some appropriate quality control processes in place.

3.5 Communicating ideas about a design model

Once a designer has formulated a 'solution' intended to meet the need of their customer, the next step is to consider *how* their ideas about the form of their design model are to be conveyed to others, since to be useful, a design needs to be implemented in some way.

Just when the concept of the creation of an artifact requiring some form of 'design activity' first emerged can only be a matter for conjecture. Many of the major creations of antiquity such as the Egyptian pyramids must have required quite a significant element of planning, so *civil engineering* in the broadest sense, may well be where this first became recognised[1]. Later on, in the thirteenth century, King Edward I of England employed 'Master James of St George' to build some of his major castles, many of which still stand today. While he was originally described as a 'master mason', his heritage leaves little doubt that he was someone who possessed significant design skills. Certainly, the nature of civil engineering did require clear directions to be conveyed from designer to the 'production team'.

When software engineering first emerged as a discipline, ideas about the form software development should take were strongly influenced by the practices used in existing branches of engineering. Hence, the notion of there being some form of distinct design step tended to be assumed in early thinking about software development, as illustrated by the idea of the 'waterfall model', shown in Figure 3.1. In this approach, the expectation is that the design model will be complete before being implemented. However, this was always regarded as rather idealised, and design activities can also be, and usually are, interleaved within the other steps of development. This is particularly useful where the details of an application are not fully known, so its design may need to evolve

[1] And Pharaoh's contract lawyers were probably not far behind in establishing themselves as a related and very profitable branch of the legal profession...

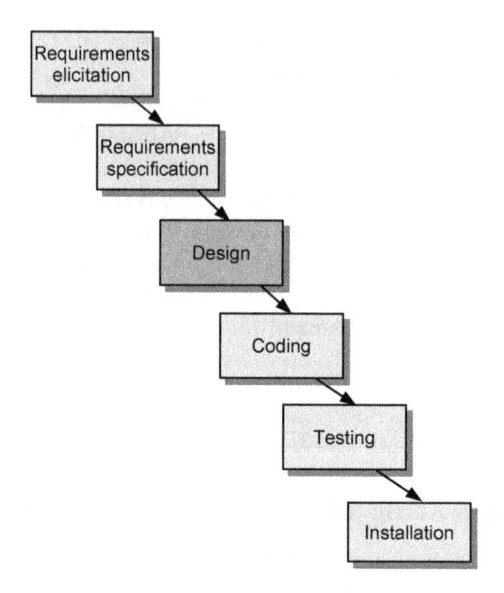

FIGURE 3.1: A waterfall model of software development

as understanding of its role emerges, leading to the evolution of more 'agile' approaches as we will see later.

So, while a design model may help the designer cope with the cognitive challenges of designing complex things, it also plays an important role in sharing ideas about the design with others. The model may be little more than a list (as in the example of moving house), or it may be much more detailed, as in the 'blueprint' traditionally produced for mechanical design. The level of detail will depend upon how much freedom can be given to those responsible for fabricating the design ideas. For mechanical engineering where end-products are not easily modified or adjusted, the degree of tolerance that can be permitted is important, so those responsible for fabrication activities will need to follow the design plan very closely. For house-movers and software engineers, whose processes can be modified rather more easily, the plan can be much less detailed, to the point that it can evolve while the implementation task is progressing. This is a point that we will return to, as it is one of the ways in which too close an adherence to traditional design thinking is often not appropriate for the software domain.

So we can conclude that *communication* with those who are responsible for fabricating the solution is likely to be an important part of a designer's role. In addition though, there is also a need for communication with those who are commissioning the design. Their requirements might be very tightly specified and the description of these may focus upon such factors as their market and other constraints ("a table for use in village halls, cheap to fabricate, durable, easily cleaned, being capable of being safely stacked, and being light enough for one person to move it easily"). Or they may be rather broad and sweeping (the medieval baron who wants "a castle bigger and grander than the one owned by their rival across the valley").

So here we see another demand upon the designer and their skill-set. The designer needs to have a good understanding of the *problem domain* (we can assume that in the examples of the previous paragraph the designer would know about tables or castles) and also of the materials that they need to work with. The great Victorian engineers like Isambard Kingdom Brunel were prepared to design railways, ships, bridges, railway stations, and many other necessary items as part of their portfolio. However, since that time design activities have tended to become more specialised, partly because of this need for deeper knowledge about the context of a design. So an aeronautical engineer is unlikely to be employed to design roads.

One thing Brunel didn't design was software!

In some ways though, software designers find themselves more in the situation of needing to create a wide range of applications, more as the Victorian engineers did (but fortunately without the need to wear tall stove-pipe hats). The use of software is so ubiquitous and its form so adaptable that over-specialisation is rarely an option. Admittedly a specialist in (say) database systems might (or almost certainly should) hesitate to take on the task of designing a real-time command and control system, but in principle at least, one person could possess the necessary range of skills for both.

So the acquisition of *domain knowledge* plays a particularly important role for software design. Some of this may be obtained from the customer's specification, but as we will see later, this may well fail to identify key elements of the 'tacit' knowledge that are familiar to everyone working in the domain—but that may not necessarily be familiar to the software designers.

If we think about designing software rather than castles, and particularly about designing software applications that will be implemented by a large team of people, what sort of thing are they likely to want to know from the designer(s)? At the very least this is likely to include:

- what the overall 'architecture' of the application is to be—what will be the major elements, what they will do and how they will interact (this will determine many things, including how the elements will behave);
- the details of key data objects that the application will need to use (usually the 'real world' things that need to be modelled or represented);
- any specific algorithms to be used to achieve the necessary functionality;

- how the elements of the application will be packaged and organised;
- how the elements of the application will interact, including any dependencies (coupling) and causal (event-driven) links;
- how the application will interface to, and interact with, its environment (users, other software components such as browsers, etc.).

Mostly these are concerned with the form or structure of the application, but some are also concerned with its behaviour, both overall and in detail. So the outcomes of the design process (the design model) is likely to be concerned with the way that the elements are *structured*, the *functions* that the system will perform, and the way that different elements will *behave*, individually and collectively. Whew!

Key take-home points about the design process

In this chapter we have considered the outcomes of the process of designing software, and addressed some ideas about what we might consider to be a 'good' design.

Abstraction is an important design tool. A designer needs to focus on the core issues that are relevant to a solution, and to successfully manage their ideas about this, needs to be able to 'discard' consideration of other issues that are not relevant to this.

Fitness for purpose should be the ultimate criterion. While a good design solution may exhibit many qualities, the ultimate need is that it should work and do what is required as well as possible.

Design models are a key product of designing. Design as a process (verb) leads to design as a product (noun). In its most simple form, where a designer also implements their design, the design model may simply reside in the designer's head, but beyond that it needs to be more formally recorded in whatever manner is most appropriate.

The rationale for design decisions should be recorded. While this may not be of immediate value to the original application development team, it will help others to ensure that future modifications and extensions will preserve the integrity of the original design structures, and so reduce the technical debt during later evolution of the system.

Design solutions need to be shared with others. While communication is particularly important where manufacturing forms a part of development, software applications are often created by teams of developers, and they all need to be aware of, and conforming to, a common set of goals and models.

Designers need domain knowledge. As well as knowing what can be done with a particular medium (in our case, software) a designer also needs to know something about the environment in which the eventual application is to operate.

Chapter 4

Design Knowledge

The previous chapters have spelt out some of the challenges associated with design in general, as well as with designing software, and from this point we focus upon the latter task. Clearly, there are people who can very ably and successfully juggle the many factors involved in designing software, and in this chapter we look at what it is that characterises expert designers, the designs that they produce, and the nature of their expert knowledge about designing.

4.1 What do expert software designers do?

When learning any skill, it is useful to analyse *what* the experts do (and *why* they do things as they do). Studying the experts might (will) not instantly turn anyone into a tournament-winning tennis player, or an award-winning artist, but it will almost certainly be a good way to improve one's skills.

In their witty little book *Software Design Decoded: 66 Ways Experts Think*, Marian Petre and André van der Hoek (2016) provide just such a study of software designers. Their book briefly illustrates and explains some of the practices that expert software designers are observed to employ. Of course this is only half of the story. It isn't just what an expert *does* that matters, it is what they *don't* do that may be just as important. (And particularly, the sort of things that novices will probably do, and that experts will avoid!)

So it is worth considering here something of what is known about expert software designers, not least because it reinforces some of the issues described in the preceding chapters. It also recognises that designing software is a *social* process as well as a technical one (Shapiro 1997). What follows is by no means everything that is known about expert designers, but we identify some significant studies (many of which have been conducted by psychologists rather than software engineers) and extract some of the key concepts and terms that emerged from these.

One of the earliest studies of how people set about the task of designing software was undertaken by Adelson & Soloway (1985). They observed how both experienced and inexperienced designers set about the task of designing solutions when the problem set before them was:

- familiar to the participant; or
- unfamiliar in detail, but set in a context/domain that was familiar to the participant; or
- unfamiliar in both senses.

They had a small number of participants, and in many ways what they termed 'experiments' might be better categorised as 'controlled observational studies'. However, some key observations that were produced from this work are ones that have also been noted in later studies of the ways that designers work, and those related to expert behaviour are summarised in Table 4.1.

TABLE 4.1: Key findings from the study by Adelson & Soloway

Observed Behaviour	**Purpose**
The designer forms an abstract *mental model* for their design.	'Mental execution' of the 'design model' assists the designer with simulating the dynamic behaviour of the eventual system.
The model is systematically expanded, keeping all elements of the design at the same level of detail.	Assists with the task of simulation, which itself becomes more detailed as the design model evolves.
Constraints on the problem/solution are made as explicit as possible.	Bounds the 'decision space' when addressing an unfamiliar problem.
Reuse of previous design plans. ('Labels for plans')	Occurs when the designer recognises that part of the design task can make use of a previous solution or part-solution. This is then 'labelled' (as a plan for how this will be resolved in the future) and the designer moves on to focus on less familiar aspects.
Making of notes about future intentions.	Aiding systematic expansion of the design model.

Simulating the 'execution' of a design model in one's head is a concept familiar to programmers, who often mentally trace through the execution of code as a check on the formulation of such constructs as conditional expressions. Arguably this is something that is particular to designing software—whereas such activities as *note-making* and *reuse* are design activities widely observed in other domains too. As we will see later, the idea of 'labelling' previously used plans is one that also resonates with the idea of the *design pattern*, described in Chapter 15.

An observational study by Bill Curtis and his co-workers examined the characteristics of a set of designers who had been identified by their organisations as being particularly successful (Curtis, Krasner & Iscoe 1988). They observed that the number of exceptional designers identified in this way was small—and hence they were usually treated as an important resource within an organisation. Interestingly though, the outstanding designers were often not considered to possess particularly good programming skills, suggesting that programming and design may call upon different abilities in an individual. Table 4.2 summarises three key findings from this study.

TABLE 4.2: Key findings from the study by Curtis et al.

Observed Behaviour	Purpose
Familiarity with the application domain.	Making it possible to map between problem characteristics and solution structures with ease.
Skill in communicating technical vision to other project members.	Much of the design work was often accomplished while interacting with others.
Identification with project performance.	Achievement of technical progress—the designer might even take on significant management responsibilities to ensure this.

In studying the ways that designers make decisions, Visser & Hoc (1990) used the term *opportunistic* to describe the way that an expert designer would revise their strategy for 'solving' a design problem. When working opportunistically, although a designer may have chosen a strategy for how they are going to develop their design ideas such as 'top-down' (systematically decomposing the solution elements into smaller actions), they may still deviate from the strategy to either:

- *postpone* a decision, perhaps because the information needed to make this has not yet been resolved; or
- process information that is readily at hand, and can be used to define design elements in *anticipation* of the way that the design will evolve.

The term 'opportunistic' was originally introduced by Hayes-Roth & Hayes-Roth (1979) in the context of planning (an activity which can be considered as a form of design, as we noted earlier). It stems from the ill-structured

Skill in communicating technical vision

aspects of design , in that, as a solution to a problem (i.e. a design model) emerges, and creates a correspondingly better understanding of the problem, this improved knowledge may well lead the designer to modify their strategy for solving the problem. So this can also be considered as a form of *feed-back* mechanism that occurs during the design process.

What else drives such opportunistic behaviour? Well, Akin (1990) has suggested that there are three 'classic' actions that occur during creative activities such as design. These are:

1. The *recognition step*, sometimes termed as the 'Aha!' response. This arises when the designer recognises a solution that has been there all along. As a creative act, it typically occurs when a designer realises that there is a simpler way of doing something than the model that they have been employing.

2. The *problem restructuring* step, in which a change in the way that a problem or design solution is being described or modelled leads to a major advance towards finding a design solution. In many ways this is similar to the idea of *refactoring* that we will discuss later in the context of agile design. As a simple example, a designer might be planning to store some of the key information for an application in a relational database, but then realises that managing the information within a set of objects will greatly simplify all of the associated operations.

3. The *development of procedural knowledge* about how a particular type of problem is best addressed, allowing the designer to perform many creative acts within a domain. One way that this is manifested in software engineering is through the use of *patterns* to provide reusable solution models.

All three actions involve recognising that a different (and better) way to address the current problem can be achieved by changing the way that it is viewed or modelled. All are related to the concept of 'design as adaptation', drawing upon the experiences of the designer as well as of others.

An observational mode of study was also employed for the project 'Studying Professional Software Designers', which is summarised in the book *Software Designers in Action* (van der Hoek & Petre 2014). Three video recordings of pairs of software designers who all addressed the same design task provided the basis for a number of different analyses, using a range of perspectives. This study also provided some of the underpinnings for the later book by Petre & van der Hoek (2016) that was mentioned at the beginning of this section.

It is useful here to look at three of the 66 characteristics of 'expertness' identified in the latter, where these identify some characteristics that have only partly emerged so far, and that are very important to the later material covered in this book. They are also ones where expert behaviour may particularly be likely to differ from the behaviour of inexperienced designers. Each characteristic has been highlighted by placing it in a box, a format that we will also use in later chapters. The numbers identify their position in the 66 characteristics. (The discussions provided here are formulated in a rather different way to those provided in their book, although they contain the same essential messages.)

Experts prefer simple solutions. PvdH #1

Experts prefer simple solutions

Given a choice between doing something in a simple way and doing it in a more complicated way, the expert will almost always choose the former. In contrast, an inexperienced designer is more likely to choose a solution that involves bells & whistles and incorporates features that 'may be useful in the future', even though (or perhaps because) their vision of what that future may be is less clear than that of the expert.

Experts externalise their thoughts. PvdH #24

What this really means is that experts *sketch*—on paper, whiteboards or any other readily available surface (paper napkins, backs of envelopes etc.). While the sketches may well be simple 'box and line' diagrams, sketching can also involve making lists, drawing tables, or a mix of the three. Externalising their ideas in this way helps the expert think them through (including use of 'mental execution' at a relatively abstract level), and adapt them. Inexperienced designers appear to be more reluctant to explore different ideas once they have a design model that they think looks feasible.

Experts externalise their thoughts

Experts invent notations. PvdH #28

This partly relates to the previous item. Essentially, what this says is that experts are not constrained to using particular notations. Part of this book is about the variety of notations used to describe software models, and about their syntax and semantics. However, while these may be (and are) useful for *documenting* and *sharing* design models, they are not always so useful for developing them, and over-attention to syntax may constrain thinking. Design sketches can take all sorts of forms (and do so) and the less experienced designer should not feel bound to keep too closely to the syntax of established notations, even when following a plan-based design strategy.

In the rest of this chapter we briefly look at the sort of experiences and mechanisms that help build up design expertise. One thing that this book definitely can't do is teach anyone *how* to be a great designer. That only comes from experience together with understanding. And so one thing that this book *can* do is to help the reader to acquire a deeper understanding of the processes involved in software design. First though, we consider a bit more fully just what the factors might be that determine how 'good' we consider a particular design model to be.

4.2 Some software design principles

When considering the design choices involved in developing a design model for a software application, what are the desirable characteristics that might make us consider a particular design model as being 'good'?

Most of us will readily accept that the software used for controlling an aircraft needs to be carefully designed and rigorously tested—particularly if we think that we might be a passenger on that aircraft one day. Yet good design matters for smaller and less obviously safety-critical things too, not least because the end-users expect efficiency (whatever that means) and reliability. We expect our software tools to behave reliably and consistently—if our bank's software systems lose an on-line transaction involving a payment into our account, it may not be life-threatening, but it can still significantly disrupt our lives. The associated *non-functional* properties are often referred to as the 'ilities', and form yet another perspective upon the complex process of addressing ISPs. And they can be much harder to assess from the design model than functionality itself (since we can assess this by use of mental execution of the model).

One way of trying to ensure that our design model will be reliable, consistent, robust, and maintainable, is to be guided by some of the basic principles that have emerged over the years. Although conformance to these cannot ensure that we have a sound design model, non-conformance will almost certainly mean that we don't have one! We outline some important ones below and will return to them at various points in the following chapters. All of them relate to the characteristics of the 'design model' itself.

4.2.1 Fitness for purpose

This should be considered as the 'ultimate' requirement for any design model—and as a principle, it is one that comes close to providing the ultimate test for an ISP. However elegant or efficient a design model may be, the two basic needs are that the eventual system should work reliably, and that it should perform the required task. That is not to say that other factors are not important, but simply that they are subordinate to the need to create a software application that does the job required of it.

Needless to say, this is an easier goal to specify than to achieve. The nature of ISPs makes it difficult to obtain a definitive specification of purpose or a test that demonstrates that it is achieved. However, what this principle really means is that the designer should not get diverted into providing unnecessary 'bells and whistles' and should focus upon the essence of what a piece of software should do. (Remember too that "experts prefer simple solutions".) The wonderful pliability of software makes it easy to add features that are not really necessary, and the temptation to do so needs to be avoided. So, what this principle gives us is a 'litmus test' for assessing a design model, in that we can look at each of its features and structures and ask if it is really essential and what it contributes to this principle.

If we ask why this should be so, the answer is very simple. Anything additional to what is needed for the core purpose of the software adds potential technical debt by incurring unnecessary complexity.

4.2.2 Separation of concerns

Most of us are familiar with the concept of the Swiss Army Knife as a tool that can be used to do many things (almost unlimited things if we have a really grand model). However, although a single knife can be used to do many things, each component of the knife has been designed to do one thing, whether it be cutting, filing, sawing etc. And this principle of having each element "do one thing and do it well" also applies to the design of software.

What this means for software is that a design model should clearly separate different elements. Input, output, data storage, handling user interaction etc. are all quite distinct tasks. And this is where the flexibility and pliability of software (Brooks 1987) can tempt the designer to cut corners and combine different things in the elements of the model. When writing software in this way, we refer to the end result as "spaghetti code", implying complexity (rather than edibility) and hence longer-term problems of technical debt, to say nothing of shorter-term problems with testing. The same issues are true for design—clear separation of concerns can greatly assist with later changes and updates to an application, as well as making for more modular testing.

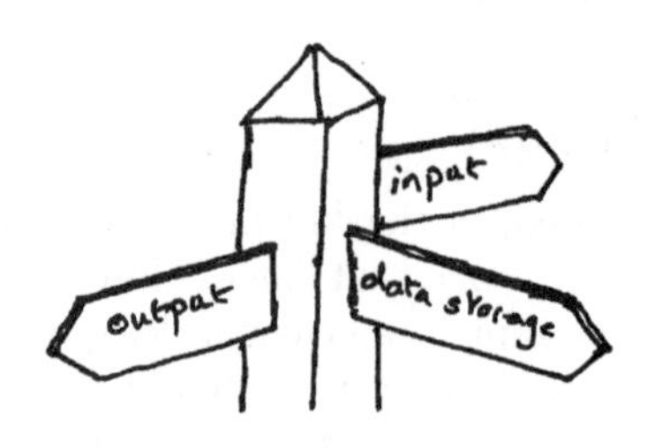

Separation of concerns

Separation of concerns in the CCC

In the context of the CCC we can identify some obvious 'concerns' that should be treated as distinct elements. These include the following (note that the list is not meant to be exhaustive).

- knowledge about the cars
- knowledge about bookings
- customer interaction
- billing for car usage
- finding available cars that are within a given area

These all interact to provide the overall functionality, but represent quite different facets of the design model.

4.2.3 Minimum coupling

When constructing anything but the smallest software applications, we need to adopt some form of modularity for its organisation, whether the modules concerned are methods (sub-programs), classes, processes or some other form. Indeed, the choice of these may well relate to the previous principle regarding the *separation of concerns*. And since these modules are part of the whole, they need to interact and share information in some manner. This inter-module dependency is usually termed *coupling*, a concept that was first recognised in the early 1970s when the modules concerned were most likely to be sub-programs (Stevens, Myers & Constantine 1974). And as new forms of modular structure for software have emerged, it is one that has continued to be relevant.

Coupling can take a range of forms. For example, module A may make use of operations, data or data types that are provided by module B. And it can occur in different ways too. Module A may invoke methods in Module B, or obtain values of variables it contains, or even inherit properties from it. The presence of any of these forms will create a dependency of A upon B, and what it particularly means is that if any changes are made to module B, they may potentially affect module A. Some coupling is of course necessary for the construction of any software system—the point is that this should be as much as is needed and no more—since unnecessary coupling can add to technical debt, complicating future evolution of an application.

Excessive or unnecessary coupling may well indicate a poor choice of modular structure. If the concerns are well separated, the dependency of a module upon other modules should be minimal, and in particular, should not involve requiring knowledge about how the other modules perform their tasks.

4.2.4 Maximum cohesion

The concept of *cohesion* is concerned with intra-module relationships. The issue here is that the elements grouped *within* a module should all share the characteristic of being related to the purpose of the module. For example, if we have a class that consists of a data structure representing some part of an application, all of the other elements of the class should be concerned with updating and reporting on that data structure. Cohesion is conceptually fairly straightforward, but can be quite difficult to assess.

4.2.5 Information hiding

This principle is also related to the way that modularity is achieved within an application, but incorporates additional notions about how information should be organised and stored within an application. It is particularly relevant to thinking about objects and in this context it is often referred to as *encapsulation*.

The basic concept is concerned with knowledge about the detailed form in which data is represented and stored. The aim is to keep this knowledge local to a module, so that it is not visible outside of a particular module (Parnas 1972, Parnas 1979). Like cohesion it is concerned with intra-module properties, and its use requires that a module should provide a number of methods that can be used to access the data, where these are used to 'hide' the way that the data is organised internally. Again, the motivation for this is to enable and simplify subsequent changes to an application.

Information hiding in the CCC

There are several items of information employed in the CCC software that should clearly be encapsulated (hidden) so that they cannot be directly modified or accessed by other objects. These include:

- details that describe the full status of a specific car—not just availability but also such things as distance covered, its current position etc.;
- the way in which information about the set of cars is organised, so that objects using this just refer to a particular car without knowing how that information is organised internally;
- information relating to customer accounts.

4.3 The evolution of design ideas

Given that we have some ideas about the characteristics of an effective and even great designer, together with some ideas about what they are trying to achieve, this then raises the question of what knowledge does a software designer need to acquire?

Long before software was thought of, people probably began learning their design skills primarily through some form of 'apprenticeship', and many would then continue to learn from sharing their experiences with their peers. There is still quite a strong element of this in many creative disciplines, and indeed, the concept of the 'design studio' has been successfully adapted for teaching about software development (Tomayko 1996).

We can reasonably assume that the craftsmen who designed the pyramids of Egypt and those who created Europe's medieval cathedrals, will have exchanged ideas with, and learned from, their peers. However, each resulting building was a single unique creation, and while new ideas did emerge, such as the use of ribbed vaulting or flying buttresses, these were more likely to relate to the design *product* rather than to the process of *learning* about designing. It was only with the emergence of what we would now consider as engineering practices that new contributions to the way that people learn about designing really began to occur. Two of these are worth noting here as being particularly relevant to software design.

1. Knowledge gained from *scientific research*. As the properties of materials became better understood, and the means of predicting their behaviour became more refined and dependable, so designers could utilise such knowledge to help create new structures. It is interesting to note that in a period when the distinctions between the professions were less rigid (and probably not even recognised) Sir Christopher Wren, the designer of some of London's most iconic churches, including St Paul's Cathedral, began his career as an astronomer and was regarded as an outstanding mathematician and geometer.

2. The concept of *reuse*. While ideas were certainly reused before the industrial revolution, the idea of fabricating structures by using 'standardised' components was not. The availability of reusable components both extends and constrains a designer's repertoire. Their use may reduce the design task as well as offering speedier development and time to market and reducing costs—while at the same time introducing new trade-offs into the design process. In particular, the opportunity to reuse things raises the question as to when it would be better to 'buy-in' existing elements and adapt one's design around these, which introduces a dependency upon a supplier, or to adopt a policy of 'build-your-own'?

   There is also the related question of how to document and catalogue components. Two of the greatest designers of the past maintained records in the form of 'sketchbooks' (Leonardo da Vinci) and the engineer's 'commonplace book' (Isambard Kingdom Brunel). These both formed an aid to their own thinking and also provided a 'pattern book' that could help convey their ideas to others.

Both scientific knowledge and reuse have had an influence upon ideas about software design as will emerge in later chapters. In the 1970s and 1980s, it was expected that using the power of mathematics could enhance the process of software design, largely by providing rigorous specification through the use of 'formal methods' (Shapiro 1997). We will examine this further later, but apart from fairly specialised areas such as safety-critical systems, mathematical forms of modelling have probably had much less impact than expected. (However, as we will see in the next chapter, scientific knowledge of a different form is playing an increasing role in software development as a whole. The development of rigorous empirical studies is increasingly providing insight into design activities.)

The concept of reuse has been much more influential and has provided software developers with an increasingly large palette of options as well as a range of forms of reusable components that have influenced design ideas and also design products. Concepts related to different forms of reuse such as patterns, software product lines and components have all enriched the design process itself, playing an important role in 'design as adaptation' alongside a designer's own experiences.

There are probably many reasons why these two practices have had rather different degrees of influence on thinking about software design. One possible reason is that they embody (or at least support) different philosophies. Mathematical forms and scientific models (as we have already observed) are essentially associated with *reductionist* thinking, commonly associated with WSPs. While that is not wholly incompatible with designing software applications, it is likely to be more successful where the requirements for such applications are very well understood. They also require an element of formal mathematical understanding for their application. On the other hand, reuse lends itself to a more *compositional* way of thinking, in which we build up a solution from parts, and hence is intrinsically more suited for use with ISPs. It certainly fits better with the idea of an opportunistic design strategy, as used by agile approaches to software development, and as such, is probably much more widely applicable to software development.

Finally, a useful concept first posited by Herb Simon is that of *satisficing*. This can be summarised as "seeking a satisfactory solution to a problem rather than the optimal one" (van Vliet & Tang 2016). Satisficing is often observed as occurring during design activities and forms an important decision-making strategy. It may be employed for a variety of reasons: a designer may have only limited time to work on an issue; there is a lack of information needed to make an optimal decision; or the designer is trying to simplify a cognitively complex issue. In essence, the use of satisficing often involves taking the first solution that 'fits' a need and can be considered 'good enough'. It may involve the use of analogy with other software applications to help make a decision, although as van Vliet and Tang observe, the risk is that "the context is different from what one is used to".

4.4 The nature of expert design knowledge

From the material of the preceding sections, we can see that a designer clearly needs to acquire quite a lot of rather complex knowledge, both about how to proceed with developing a design, and also about how to model their ideas. Since much of the content of Parts II and III is concerned with describing ways of sharing design knowledge and ideas, it is important to understand something of the challenges such sharing poses, and the associated limitations in what can be done.

Indeed,while the idea of being able to transfer design knowledge from experts to less experienced designers is an attractive one, when we look at the nature of design knowledge, the reality involved in doing it is rather more complex. To begin with, an expert designer may well be unable to explain *why* they know how to do something. As an example, someone who is an expert in graphic design may know which typefaces can be most effectively used for a particular type of document, but be unable to provide a rationalised argument as to why this combination works well.

Such design heuristics or 'rules of thumb' are no less effective or less valid for the lack of an explanation. Indeed the tacit knowledge that underpins them may be based upon such deep and complex cognitive issues that any explanation will be of limited general usefulness. When studying how people design software, Vessey & Conger (1994) observed that "examining expert problem solving can be quite difficult, since experts automate their problem-solving processes to the point at which they are no longer able to articulate what they are doing". Unfortunately, without some understanding of *why* particular solution strategies work—one of the characteristics that distinguishes an engineering discipline from a craft—a heuristic may be limited in usefulness. In a domain where the problems are relatively repetitive (something that characterises many crafts), this may not be too critical an issue. However, in software development, although there may be common elements between different applications, the requirements that they address are usually quite distinctive.

An obvious, and related, question is how do designers 'store' their expertise in their brains? While this is really a question for which any serious answer must fall well beyond the scope of this book, there has been research that examines how software designers organise their design knowledge. Clearly this is not something that can readily be observed, but in Détienne (2002), the author uses the term *schema* to describe an important and rather abstract form of design knowledge. There are different forms of schema, and in this model, programming expertise is partly based upon the possession of a set of *constructional schemas*, which may be further reinforced by a set of *domain-specific schemas* that help map and adapt the programming schemas on to certain types of problems. So implicitly, in addressing a design problem, an expert may well make use of a number of such schemas, where these address different facets of the problem.

Experts develop knowledge schemas

Not only does the transfer of knowledge pose a challenge, the *quality* of much of our knowledge about how to design software may also be rather mixed. The concept of *shells of knowledge* proposed by Brooks (1988), provides a useful way of categorising the confidence that we might be able to place in specific pieces of design knowledge, using the following three classifications.

1. *Findings*, or well-established scientific truths, with their value being measured in forms that correspond to the rigour with which they have been established. In terms of empirical software engineering concepts, these would most likely correspond to findings from a systematic review (Kitchenham, Budgen & Brereton 2015).

2. *Observations*, that report on actual phenomena, with their value being determined by the degree of 'interestedness' that they possess for us. Again, in terms of empirical studies, these would be the outcomes from a 'good' primary study, whether experimental or observational in nature.

3. *Rules of Thumb*, which consist of generalisations that are 'signed' by the author, but not necessarily supported by data, with their value being determined by their usefulness. These might well be found in 'experience reports'—lacking empirical rigour, but possibly providing useful, if informal, analysis of experiences.

Advances in empirical software engineering have improved our knowledge about many aspects of software engineering in the period since Brooks proposed this model, and these are reviewed in the next chapter. In a paper by le Goues, Jaspan, Ozkaya, Shaw & Stolee (2018) the authors offer a more detailed set of categories for the quality of evidence and discuss how these map on to this smaller set proposed by Brooks. However, regardless of how we categorise software engineering knowledge, the quality of much of our knowledge about designing software largely falls into the categories of *observations* and *rules of thumb*.

So, given that design knowledge is difficult to codify, and its quality is mixed, this emphasises why expert knowledge about how to design software systems can be difficult to share with others. From the 1970s onwards, various mechanisms have been developed to assist with this, starting with plan-driven design methods, and going on through such ideas as software architecture, agile methods and design patterns. In essence, these provide ways of translating expert design schemas into some 'rules' that can be used when designing software intended to meet specific purposes. While they may be relatively crude, relative to the subtlety of the cognitive knowledge involved, they have nonetheless shown themselves to be useful, and we will be looking more closely at them in Parts II and III of this book.

Key take-home points about design knowledge

This chapter has examined some of the characteristics of expert software designers and of the ways that they design software systems.

Experienced designers use abstract mental models. An important role for these models is to allow the designer to mentally 'simulate' the way that a system will behave in particular conditions. This characteristic is one that is largely peculiar to software design, since the 'product+process' nature of software forms one of its distinctive features.

Designers often work opportunistically. Rather than following some fixed set of procedures to develop their design model (as advocated by plan-driven design approaches), experienced designers will employ a general strategy but adapt it as their understanding of the problem (and solution) evolves.

Designers sketch their ideas. Design ideas are fluid and experts like to concentrate on abstractions which can easily be revised to allow ready exploration of their models for a solution. Using informal notations to aid the development of a design model helps with this by avoiding the constraints of a fixed syntax and semantics.

Design principles embody accumulated design experience and provide a set of criteria for assessing the quality of a design model.

Designers reuse. This may occur at the abstract level, where designers reuse part-solutions that they previously found to be successful, and also at the level of reusing actual software components that are known to work and be trusted, restructuring and adapting their design plans to accommodate these where appropriate.

Design decisions may rely upon *satisficing*—whereby the designer opts to use a 'good enough' solution to a need, rather than expending time and effort to try and find an optimum solution.

It may be difficult to articulate and codify design knowledge. So far, everything described in this book emphasises that design cannot be performed by 'following a procedure'. Indeed, successful designers may find it difficult to explain why they make particular decisions, since their expertise is encapsulated in the form of a set of *schemas* rather than procedural rules.

Chapter 5

Empirical Knowledge about Software Design

A key characteristic of this book is that wherever possible, when discussing issues about software design, the reader is provided with relevant empirical *evidence* to support the discussion. So this chapter provides a brief outline of what typically constitutes such evidence and how it is acquired.

We begin by discussing some issues related to measurement, since this underpins any discussion of empirical studies. We then go on to discuss something of these studies, what we mean by evidence in this context, and how it might be used to support software design.

What we will see is that acquiring empirical knowledge about software design issues can be quite challenging, and may not always lend itself to providing detailed guidance to the designer. However, empirical studies can often provide considerable insight into design choices, while also forming a more rigorous source of knowledge than 'expert opinion'.

5.1 Measuring software development processes

Empirical knowledge is usually based upon the use of some form of measurement. So, before discussing the forms of empirical study and the types of knowledge that they can provide, it is important to understand the forms of measure commonly used in empirical studies.

Many of our ideas about what constitutes 'measurement' stem from disciplines such as physics that involve measuring the properties of real-world phenomena. An important feature of subjects such as physics (as well as most other sciences and technologies) is that they are founded upon a *positivist* philosophy, embodying the assumption that there are general rules that apply to the phenomena being studied. In the physical world that philosophy is commonly manifested in such assumptions as "all electrons have the same mass and charge"[1]. A further assumption is that the 'observer' who is performing the measurements has no influence upon the values obtained.

Both of these assumptions present a challenge when we seek to study *human-centric* activities such as designing software. For practical purposes we have to adopt a positivist philosophy, while accepting that this is probably something of an approximation given that people have different abilities. And when designing our studies, we have to plan them so as to minimise the extent to which bias can arise from the involvement of the observer, particularly where any observations may involve an element of *interpretation*.

5.1.1 Measuring physical phenomena

In the physics laboratory, a desirable form of experiment is one where the observer has as little to do with the measurement process as possible. If the measurement can be performed by instruments that (say) count the number of electrons hitting a 'target', then although measurement errors cannot be wholly eliminated, they can usually be kept to a very low level.

Often however, it isn't possible to avoid having the involvement of an observer when making measurements. For example, analysing traces that show how much a beam of electrons is deflected by a magnetic field. For this, some errors are likely to occur because the observer didn't have their head in quite the right position to read the relevant point on the scale, or simply made a small error in identifying or recording what the value should be.

Scientists are accustomed to this form of error and its properties. For one thing it is expected that over a large number of measurements there will be roughly as many 'over-measures' as 'under-measures', and that errors of this type will have a *normal* distribution, based on the well-known bell-shaped curve. Experimenters will be expected to make an assessment of the magnitude of such errors and to report them. While this will represent a (usually small) degree of uncertainty about the measured value, the important thing is that this uncertainty essentially stems from the limitations of the measurement instruments. Taken together with the original assumption that there is a single

[1]The opposite philosophy is termed an *interpretivist* one. Interpretivists view each measurement as being unique, so that it can only be understood in terms of its specific context. While this perspective is one that is adopted by many social scientists, research into software engineering practices usually adopts a positivist approach.

'right' value, the expectation will be that any variation that we see in the measured values can be attributed to the way they were obtained.

5.1.2 Measuring human reactions

Humans (a category generally accepted as including software developers) introduce a whole new element of complexity when we seek to make measurements related to the consequences arising from their actions.

Human-centric studies take many forms, as we will see in the next section. Usually they involve studying the effects arising from introducing some form of *treatment* or *intervention* to some situation. This requires the collection of measurements for one or more *dependent* variables in order to assess the effects. To explain about the issue of *measurement*, we focus on two major categories of treatment.

- The first is those studies where the human participant is a *recipient* of the experimental treatment. An important form used for these is the *Randomised Controlled Trial* (RCT), as used in many branches of medical research. An RCT can be used to make comparison between the effects arising from the application of some form of experimental *treatment* and what happens when no treatment (or possibly, another treatment) is applied. An RCT uses two groups of human participants. One is the *experimental* group, whose members receive the treatment being investigated, in whatever form this might be administered, such as the use of tablets. The second is the *control* group whose members receive what appears to be the same treatment, but where this either employs some form of harmless *placebo* or consists of using an existing treatment for comparison.

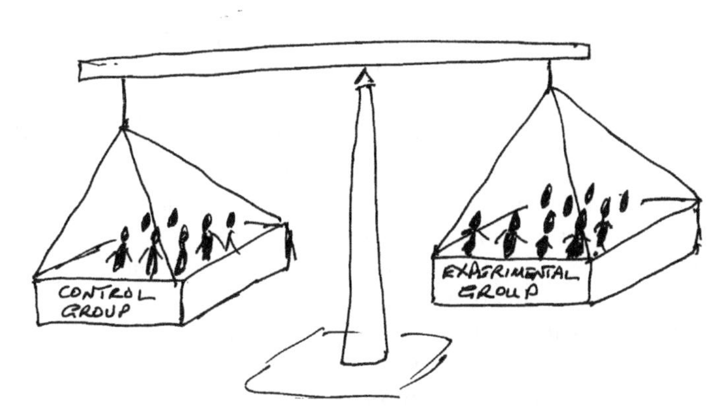

An RCT compares the effects experienced by two groups

The key issue here is that the experimenters expect to observe considerable variation when measuring the reactions of the participants in the two groups. This is because humans are complex and different and while everyone in the trial will have been selected because they have some condition that could potentially be affected by the treatment, this may occur in different degrees, and they may also have other conditions, possibly unknown, that could affect the outcomes.

Some people in the experimental group may therefore respond well to the treatment, but others less so, or even negatively. As a result the measured outcomes will vary between participants, and to a much greater degree than would arise from measurement errors alone.

So it can't be assumed that any measurements across each group will be spread in a normal distribution. This makes it necessary to use reasonably large groups of participants for such studies, together with statistical tests to determine whether or not the spread of results arises from the treatment, or could have been produced at random (that is, with the treatment not being the cause of any observed effects).

- The second category contains those studies where the treatment involves the participants in *performing* some form of activity. This is of course what arises in software development, where the activities may involve using a particular test strategy to debug a block of code, or revising a design that was produced using a specific technique. The use of RCTs is clearly impractical for this type of study, since participants can hardly be unaware of which group they are in. Although experiments and quasi-experiments[2] can sometimes be used to make comparisons, it may well be difficult to identify a suitable form of activity for the control group. So in order to study such activities we often end up using *observational* forms of study. (The study of designer behaviour by Adelson & Soloway (1985) described in Chapter 4 is an example of an observational study.)

Since everyone taking part in such a study comes along with their own experience and opinions, together with their natural ability, it can be expected that there will again be significant variation in the values for any outcome measurements of the way that the participants react to the intervention (or its absence), such as the time to complete a task, or the sort of 'solutions' produced. Again though, measurement of what appears to be a relatively simple choice of dependent variable (the time to complete a task) can be confounded by the ill-structured nature of such tasks, since it may well be difficult to define exactly what constitutes 'completion'.

When we do perform some form of experimental study, we often plot the results using something like the *box plot* illustrated in Figure 5.1. This shows

[2]A *quasi-experiment* is a form of experiment where some factor cannot be randomised across the participants (Shadish, Cook & Campbell 2002). This form is particularly relevant where the factor might be something like experience of performing object-oriented design.

both the distribution of the results, as well as such aspects as 'skew' (where results are biased to the positive or negative side). And although it appears that, with a few exceptions, most people perform the task more quickly using the experimental tool, the task of determining whether this means that we are seeing a real effect or not still has to depend upon a statistical analysis.

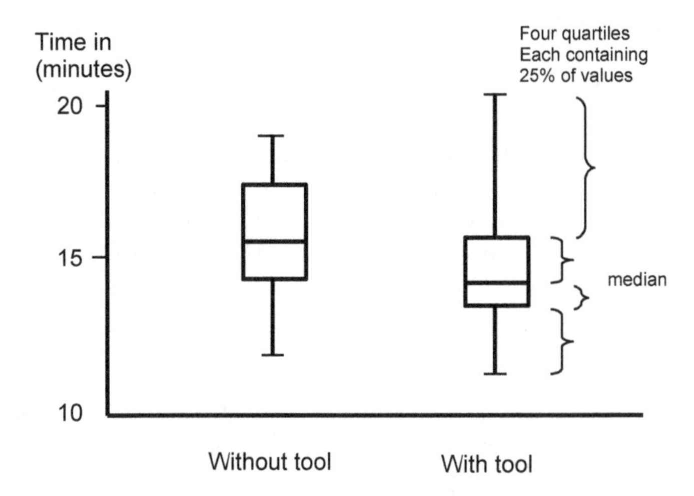

FIGURE 5.1: Using a boxplot to summarise the outcomes of a study

Why do we need to know about these things in a book about software design? Well, firstly to understand that conducting empirical studies that investigate design activity are by no means simple to perform. nor are the properties that we are interested in necessarily easy to measure. Secondly, to appreciate that the naturally-occurring variations that arise in the measured values will mean that any findings from such studies are unlikely to be clear-cut. That isn't to say that there won't be any useful findings, but rather that the findings will be subject to some sort of 'confidence' assessment, based upon the extent to which those conducting the study can be sure that the effects that they have observed stem from the 'treatment' employed in the study, and not from other *confounding factors* that are not controlled by the investigators.

5.2 Empirical studies in software engineering

When software engineering first emerged as a distinct sub-discipline of computing, empirical knowledge was largely acquired from experiences gained during the creation of software applications 'in the field', and was codified by identifying those practices that seemed to work well. This was assisted by the way that many software applications used a limited set of architectural forms for their implementation. Writing down guidelines so that others could benefit from these experiences was therefore unquestionably useful, but such *experience reports* did lack rigour and might fail to record and report all of the relevant factors.

Through the 1990s, there was a growing interest in performing more rigorous empirical studies to investigate how software engineers performed a range of tasks. Much of this was concerned with trying to understand how they acquired and used the knowledge needed to underpin their activities and decisions. This might seem strange, as the obvious comment is "didn't they know that". However, as noted in the previous chapter, expert decision-making draws upon many forms of knowledge, much of which may be tacit, to the point where an expert cannot explain the exact reasons for making a particular choice.

5.2.1 The empirical spectrum

Empirical studies take a wide range of forms. However, for our purpose we can identify two important groups of studies as follows.

- Studies of individuals that try to elicit the basis on which an individual makes decisions (that is, *why* they make them as they do). Primarily these are *qualitative* studies that involve making detailed observations, and are usually conducted with a small number of experts as the participants. Analysis of qualitative studies is likely to involve categorisation, narrative reporting, and interpretation of what the participants say. Such studies are also likely to be *field studies* in which any observations are made in the workplace when the participants are working normally.

- Comparisons between groups of participants where the members of a group have (preferably) similar levels of expertise, possibly for the purpose of comparing expert and novice behaviours. Often these are *quantitative* studies, drawing upon the positivist assumption that many individuals will behave in a similar manner when faced with a particular problem. Data collection will involve some form of counting and analysis is likely to employ statistical techniques. Quantitative studies are likely to be more concerned with *what* is done than with *why* it is done. Such studies are more likely to take place in the *laboratory*, where conditions can be controlled. However, this does introduce an artificial element to the context, and usually limits the duration of a study.

Empirical software engineering draws upon various other disciplines for its research practices, adapting them as necessary (Kitchenham et al. 2015). In this section, we focus upon those characteristics that will help us understand what the outcomes from different studies might mean.

5.2.2 The research protocol

Empirical studies in software engineering are commonly human-centric, studying how software developers undertake the various activities involved in creating and updating software systems. An important role in this type of study is that of the *observer*, who is often called upon to make interpretations of what they observe, not least because participants do not always behave quite as anticipated.

One of the hazards of such interpretations is that, since the observer will be aware of what the study is investigating, they may be unconsciously biased or influenced by this when making their interpretations. To help avoid, or at least reduce, the degree of any resulting bias, it is common practice to develop a comprehensive plan for conducting and analysing an empirical study, which is known as the *research protocol*.

The research protocol may well be quite a substantial and detailed document, but inevitably it may still not anticipate every situation. For example, when faced with a particular choice as part of a study, the protocol might expect that participants will usually make one of three fairly obvious choices, and so the task of the analyst will be to determine how to categorise each participant's actions against these. However, when the study is performed, it may well be that a small number of participants make another, and significantly different, choice. In that case, the research team may agree to create a fourth category for purposes of analysis, and this decision is then recorded and reported as a *deviation* from the protocol. Deviations may occur for all sorts of reasons, including errors that the experimenters make in conducting the study, but may also occur as a result of encountering unanticipated participant actions.

So the purpose of the research protocol is to help ensure that the study is conducted rigorously, and that in analysing the results, the experimenters do not 'fish' for interesting results.

5.2.3 Qualitative studies

Qualitative studies are apt to be observational in their nature and concentrate upon gaining insight into the reasoning that lies behind the activities of an individual designer. Here we provide two examples of the ways that such studies might be conducted (they are by no means the only ones).

- *Interviews*. One way of finding out what designers think about what they do (and why they do it that way) is to talk with them. An interview lets the researcher probe into issues that look promising, but equally it needs to be kept under control. A commonly-used procedure for organising this is the *semi-structured interview*, in which the interviewer has a set of questions developed in advance, but can deviate from these to pursue issues that arise, as and when necessary.

- *Think-aloud.* This is really one element in the technique known as *Protocol Analysis* developed for knowledge elicitation in psychological studies (Ericsson & Simon 1993). It involves asking the participant to undertake some task and to describe their thinking verbally as they perform it (*think-aloud*). A transcript of their verbal 'utterances' is then analysed, those utterances that are not directly relevant to the task are discarded, and a model is developed around the others. In practice, think-aloud is by no means easy to perform (among other problems, participants tend to 'dry up' when they are concentrating), but it can provide valuable insight into the basis for their actions, including identifying where this involves the use of tacit knowledge (Owen, Budgen & Brereton 2006).

A challenge when conducting qualitative studies is to be able to analyse the outcomes in as rigorous and unbiased a manner as possible, particularly since an element of human interpretation is usually required. To help reduce the risk of bias, good practice involves using more than one person to analyse the data, with the analysts working independently and then checking for consistency of interpretation between them. And of course, the analysis should be planned in advance as part of the research protocol.

We will examine a number of outcomes derived from such studies as we progress through the following chapters, and some of the studies described in the previous chapter such as (Curtis et al. 1988) employed qualitative forms.

Qualitative studies tend to use *ordinal* or 'ranking' forms of measurement scale. So, when making measurements, the observer (or participants) will typically rank (say) issues or preferences in terms of their perceived importance, with no sense that the elements are equally spaced.

5.2.4 Quantitative studies

Because quantitative studies are more likely to be laboratory-based and use statistical techniques for analysis, they usually need larger numbers of participants than qualitative studies, in order to provide an appropriate level of confidence in the results. Two of the more widely-used tools of quantitative studies are experiments (and quasi-experiments), and surveys.

- *Experiments* and *quasi-experiments* take a range of forms, but are usually *comparative* in form. The purpose is to test a *hypothesis* and to establish where this is true or not. More precisely, the aim is to establish whether the hypothesis or the *alternate hypothesis* is true. For example, a hypothesis to be tested might be that design models created by using the *observer* design pattern (described in Chapter 15) will produce larger classes than those that do not use patterns. The alternate (or *null*) hypothesis will then be that there will be no differences in terms of the sizes of class involved.

  Experiments are similar to RCTs, but less rigorous, because the participants cannot be blinded about their role. Most experiments performed on software engineering topics are actually quasi-experiments. This is because participants can rarely be randomly assigned to groups, as this is often done on the basis of experience (Sjøberg, Hannay, Hansen, Kampenes, Karahasanovic, Liborg & Rekdal 2005).

As an example, an experiment may be used to investigate whether inexperienced developers can make specific modifications to code more easily when the code has been structured using a particular design pattern. To reduce any bias that might arise from the choice of the experimental material, each participant may be asked to make changes to several different pieces of code, although this then complicates the analysis!

- *Surveys* provide what may be a more familiar tool. However, it can be difficult to recruit enough participants who have relevant expertise. For example, if a survey seeks to ask participants who have appropriate expertise with object-oriented design to rank a set of different object inheritance notations, then recruiting a sufficiently large set of participants who have the time and inclination to do this will be difficult, as will determining just what comprises 'sufficient expertise'. As a result, the design and conduct of a survey can be just as challenging as designing an experiment (Fink 2003, Zhang & Budgen 2013).

As indicated in the previous section, there are many factors that may influence individual choices and decisions, and when studying software design, one limitation with using quantitative forms is that they provide little explanation of *why* particular decisions are made. Their positivist nature also assumes that a degree of commonality in reasoning lies behind the particular decisions made by the participants, which in many ways conflicts with the whole ethos of designing something being considered as an exercise in individual skill.

Quantitative studies usually employ 'counting' forms of measurement scale (*ratio measurements*). In the example of a survey given above, this might involve counting how many times each notation appeared as the top choice for a participant.

5.2.5 Case studies

In recent years, the *case study*, long regarded as a major research tool for the social sciences, has become more widely used in software engineering (Runeson, Höst, Rainer & Regnell 2012, Budgen, Brereton, Williams & Drummond 2018). However, the term is sometimes used by authors in a rather informal way, so some care is needed to determine the degree of rigour involved in an individual case study.

A case study can be viewed as a structured form of observational study. When conducting an experiment the researcher aims to get many repeated measures of the same property. For example, the amount of time participants need to complete a design task when using a particular strategy. Since we can expect that this will vary considerably for different people, we need to measure the time for many participants. In contrast, a case study seeks to measure many different variables related to a particular 'case', which might be a software development project, and then to *triangulate* the knowledge provided by the different measures to see how well they provide mutual reinforcement about some particular characteristic.

Triangulation aggregates different measures

While many social scientists consider case studies to be an *interpretivist* form of study, with any findings only applying to that particular case, others do consider that the findings of different case studies may be combined, so providing a *positivist* interpretation. Robert K Yin has been a major proponent of the positivist approach to case study research, and the use of case studies in software engineering has largely been based upon his work (Yin 2014).

Yin suggests that case studies can be particularly useful when addressing three types of research question, as described below.

- An *explanatory* case study is concerned with examining *how* something works or is performed, and identifying any conditions that determine *why* it is successful or otherwise.

- A *descriptive* case study is more concerned with performing a detailed observation of the case and therefore "providing a rich and detailed analysis of a phenomenon and its context". Essentially it is concerned with identifying *what* occurs.

- An *exploratory* case study forms a preliminary investigation of some effect, laying the groundwork for a later (fuller) study of some form.

Case studies can use a mix of qualitative and quantitative data, and both explanatory and descriptive case studies may usefully make use of one or more *propositions* that direct attention to some aspect that should be investigated.

Propositions can be considered to be a less formal equivalent to the *hypothesis* used to pose the research question for an experiment.

Case studies are a particularly useful way to conduct *field studies* that involve experienced industrial practitioners working in their own environment, rather that in a laboratory context (as occurs with experiments). Their use overcomes the practical difficulty involved in having enough practitioners available to perform an experiment, and makes it possible to conduct larger scale investigations, often over a relatively long period of time ('longitudinal' studies). An analysis of the 'primary' studies used in a set of systematic reviews in software engineering over the period 2004-2015, suggests that many of these can be classified as case studies (Budgen, Brereton, Williams & Drummond 2018), although many authors did not give a clear definition for exactly what they classified as being a case study.

5.3 Systematic reviews

The forms of study described so far are all examples of what we term *primary studies*. Primary studies directly study the issue of interest in a particular context. This section examines the use of *systematic reviews*, sometimes termed systematic literature reviews, which are a form of *secondary study* that seeks to locate all of the 'primary studies' that are relevant to its research question and then aggregates and synthesises their results in order to provide a more comprehensive answer to that question.

A systematic review synthesises the findings of many primary studies

Systematic reviews provide an increasingly important source of knowledge about software engineering practices. The concept of the systematic review was originally developed in the domain of clinical medicine, where it forms a major source of clinical knowledge for a wide range of healthcare interventions and the basis for what is termed *Evidence-Based Medicine* (EBM). They have since been adopted and adapted for a whole range of other disciplines, including *Evidence-Based Software Engineering* (EBSE) (Kitchenham et al. 2015).

The procedures followed in performing a systematic review are intended to reduce the variation in the outcomes and produce findings that are less susceptible to bias than those from individual studies. The 'systematic' aspect is especially important when planning and performing such a review (and like all empirical studies, a systematic review requires a detailed research protocol).

Two elements where the systematic aspect plays a major role are in the process of *searching* for primary studies, and also when performing the activity of *inclusion/exclusion* that determines which primary studies will be included in the review on the basis of their relevance and quality. Performing these tasks systematically helps to overcome one of the problems of 'expert reviews', in which an expert synthesises the primary studies that they think are relevant to a topic (which may well be those that agree with their opinions of course).

As might be expected, there are differences between the use of systematic reviews in software engineering and their role in clinical studies. The use of RCTs for clinical studies, and the role of participants in these as recipients of a treatment, make it possible to use statistical techniques such as meta-analysis to synthesise the outcomes of multiple studies. This strengthens the confidence that can be placed in the findings of a systematic review. In software engineering, systematic reviews often have to use rather weaker forms of synthesis to combine a range of primary study forms (Budgen, Brereton, Drummond & Williams 2018), making the findings correspondingly less authoritative.

When used in software engineering, systematic reviews can involve synthesising both quantitative and qualitative forms of primary study. When used to synthesise quantitative studies, they may be employed in a role that is somewhat closer to how they are used in clinical medicine, being used to address research questions that are comparative in form (such as 'does use of treatment A result in fewer errors per object than the use of treatment B?'). Whereas systematic reviews that involve synthesising qualitative studies tend to address more open research questions such as 'what are the problems encountered in adopting practice X within a company?'.

In clinical medicine, education, and other disciplines that employ systematic reviews, these are apt to be commissioned by government departments, health agencies and the like, to help provide advice about practice and policy. However, in software engineering they have so far largely been driven by the interests of researchers. One consequence is that they are less likely to provide findings that are relevant to, and directly usable by, practitioners. A study seeking reviews with outcomes that were useful for teaching and practice found only 49 systematic reviews published up to the end of 2015 that had 'usable' findings, from 276 reviews that were analysed. (Most systematic reviews were concerned with analysing research trends rather than practice.) However, as the use of systematic reviews in software engineering matures, this should begin to change (Budgen, Brereton, Williams & Drummond 2020).

The important thing about empirical studies in general, and systematic reviews in particular, is that not only can they provide more rigorously derived empirical knowledge about software engineering practices, they can also help would-be adopters of these practices to assess how useful the practices may be to them.

5.4 Using empirical knowledge

Each of the chapters following this one has a short section discussing the available empirical evidence related to the topic of the chapter. (Well, Chapter 18 doesn't, but there is a discussion in each section instead!) Most of this evidence comes from the findings of systematic reviews.

Where a systematic review performed in software engineering provides findings that can be used to advise practice, the empirical knowledge derived from its findings is often presented in one of three forms (or a combination of these).

1. Knowledge that has been derived from the *experiences* of others, usually in the form of 'lessons' related to what works (or doesn't) when doing something. A good example of this is the benefits of reuse in industry studies by Mohagheghi & Conradi (2007).

2. A list of *factors* to consider when using some practice (a more structured form of experience). Again, a good illustration of this is the study of pair programming in agile development by Hannay, Dybå, Arisholm & Sjøberg (2009).

3. Ranking of knowledge about different options to help with making *choices* between them. The study of aspect-oriented programming by Ali, Babar, Chen & Stol (2010) provides an example of this form of knowledge classification.

All three forms are relevant to studies of design, as demonstrated by the examples, although systematic reviews that explicitly address design topics are still relatively uncommon. There may be several reasons for this, but one is certainly the shortage of primary studies (Zhang & Budgen 2012). However, it is worth noting that many practically-oriented systematic reviews in software engineering do tend to make the most of the available material synthesising the findings from a variety of forms of primary study, including case studies.

An interesting analysis of the use of evidence in another discipline (management research) by Joan van Aken (2004) makes a useful distinction between the ways that research knowledge is used in different disciplines.

- *Description-driven research* is commonly used in the 'explanatory' sciences, such as the natural sciences, for which the purpose is to "describe, explain and possibly predict observable phenomena within its field".

- *Design sciences research* such as performed in engineering, medical science and other health-related disciplines, for which the purpose is to "develop valid and reliable knowledge to be used in designing solutions to problems".

So, where does the use of systematic reviews in software engineering research fit into these categories? Well, researchers would probably view their use as performing a form of description-driven research, while practitioners would probably prefer to be provided with findings from design sciences research. For the purpose of this book, it is definitely the latter group that is of interest.

Of the 276 systematic reviews analysed in Budgen et al. (2020), the bulk, described earlier as being focused on studying research trends, really fit into the first category and can be considered as largely description-driven research. The 49 that were considered to have findings relevant to teaching and practice probably mostly fit into the second category of design sciences research, in the sense that they provide some guidance for problem-solving. And as new studies emerge there should be a growing corpus of useful guidance derived from empirical knowledge.

Key take-home points about empirical knowledge

This chapter has introduced concepts about measurement and empirical studies that are important when seeking to understand the nature and form of design knowledge that can be obtained through the use of empirical research.

Variation is to be expected in the measurements that are obtained when performing the sort of human-centric studies used to answer questions about software engineering and its practice. This variation is natural, and arises because each participant draws upon individual experience and different levels of expertise when performing the experimental tasks.

Empirical studies of software design activities are challenging to perform, and so any findings should be associated with some level of 'confidence', based on the procedures followed in conducting the study.

The research protocol is an essential element when undertaking a rigorous empirical study. The protocol should incorporate plans for performing the study and analysing the results, and should be produced in advance of undertaking the study.

Both quantitative and qualitative studies have value for research into software design issues. They respectively address questions related to *what* occurs and *why* something occurs.

The use of systematic reviews can help overcome the effects of bias that can arise in the findings of individual 'primary' studies and hence can provide the most reliable sources of evidence for making design decisions or selecting a design strategy. The empirical knowledge provided by such evidence can be presented for use in a number of ways, including: summary of experience; lists of factors to consider; and comparative ranking of techniques.

Part II

Design as a Noun: How Software Is Structured

Chapter 6

Software Architecture

The idea that a software application might have an *architecture* has become an important concept in thinking about how applications are composed, and how their operations are organised. As in other domains, its value largely stems from providing a highly abstract and readily visualised perspective on how something is organised and structured, whether it be a building, a ship, or a software application. And the idea of architecture also captures more than the way something is constructed. It also embodies ideas about how it functions (think of how 'tanker', 'car ferry' relate to a combination of different forms and functions when considering naval architecture).

So, *architecture* turns out to be a versatile concept: we can describe an application as having 'an architecture'; use ideas about 'architectural style' to categorise architectures; and employ 'architectural patterns' when designing an application. These facets are all explored in the rest of this chapter, but first we look a bit more closely at what we mean by the concept itself.

6.1 What architecture provides for us

When software applications consisted of programs executed on isolated computers (the 'pre-internet era'), there was little need for the concept of architecture. Software was largely developed using imperative programming languages (Algol, COBOL, FORTRAN etc.) which used broadly similar structures. Certainly, there were different ways that an application could be organised. It might be structured as a single process with a 'main' program invoking sub-programs; or as a database, organised around tables of data; and it might be organised as a sequence of connected processes (through such mechanisms as the Unix 'pipe' that provided data-streams for this purpose). An application might even run as a set of concurrent processes interacting across a small local network of computers that shared a common area of memory. But only a limited vocabulary was needed for describing these distinctions.

Various influences played a part in subsequently extending the options available to the software developer. Some arose from developments in implementation technology (objects, services), while others were concerned with the way that the internet introduced new ways for an application to interact with the world. And as these combined and proliferated, so too did the range of forms that software could take, and hence the need to codify the role of 'architecture'—recognised in the pioneering work of Shaw & Garlan (1996).

Software components interact in many different ways (invocation, uses, inheritance, data transfer etc.). All of these interactions create dependencies or *coupling* between the components, and this remains as important a concept now as it was when first identified—not least because the range of forms it can take has expanded.

The implication of concepts such as coupling and that of *separation of concerns*, is that we should seek to make the major parts of our application as independent of one another as possible, both when considering functionality and also information. And it is in finding ways to provide such independence that the choice of an architecture plays a useful role. When designing the architecture of an application we need to devise ways of keeping coupling to a minimum, and the manner in which this should be organised should obviously depend upon the task that is to be performed by the application. So, the way that (say) the components of a compiler are organised, will need to be quite different to the way that we might organise the components of a word processor, or those of an on-line multi-user game.

Other factors may also influence the specific choice of architectural form for an application. One of the earliest discussions of the benefits of thinking of the organisation of an application in terms of architecture was provided in Garlan & Perry (1995). Introducing a collection of papers on software architecture, they described many of the key issues lucidly and concisely. In particular, they argued that the concept of architecture can have a positive impact upon each of the following aspects of software development.

Understanding. Architectural concepts provide a *vocabulary* that can aid our understanding of an application's high-level design (model).

Reuse. The cost of developing software encourages the *reuse* of existing code (and designs) wherever possible, and the high-level components comprising a system architecture may assist with identifying where there is scope for reuse.

Evolution. When designing a software application, we need to consider how it might evolve, and facilitate future changes by allocating the elements that we think most likely to change into distinct sub-systems. Doing so also provides a guide for anyone who may later need to change the design of the application—ensuring that any changes conform to the architecture should help retain the overall integrity of the application.

Analysis. The architecture can help with *checking* for consistency across different characteristics of a design.

Management. Anticipating how an application may evolve is (or should be) a factor in determining its architectural form. Implicitly, making a poor choice can create a technical debt that will constrain future evolution.

The last point captures an important aspect of architecture. For any application, the choice of its architecture is an important and *early* decision, influencing the later design stages as well as implementation. And not only is this decision made at an early stage in development, it is also one that is difficult to change significantly at a later time since the choice of architecture underpins so many aspects. So thinking about how a system will evolve might be something that is difficult to do, but it is something that is vitally important and that needs to occur right at the outset of the design process (van Vliet & Tang 2016).

Thus far we have confined discussion of architecture mostly to some fairly abstract ideas of what an architecture is and what influence it has. The next step is to consider how we might describe the form of an architecture, and that requires that we employ the idea of an *architectural style*.

6.2 Architectural style

The concept of *architectural style* is both useful and something that is familiar from other domains. When we describe a building to someone else, we might refer to it as being in a 'Tudor style' or a '1960s style'. For buildings, descriptions of their style may also be related to the materials used in their construction. So, although we often use such a temporal label to describe a style (and implicitly the materials used in their construction), we also use terms like 'New England' or 'Steel Frame', reflecting the way that the form of buildings has tended to evolve as new materials for their construction came into use. And as already noted above, in the context of 'naval architecture' we tend to use labels that relate to function, such as 'oil tanker', 'car ferry' etc. For both buildings and ships, the use of such a label gives us an idea of the characteristics that a particular building or ship might have.

For software, the concept of architectural style can similarly be viewed as providing a very abstract description of a set of general characteristics related to the form of an application. When provided with such a description, we will then know something about the general form this has, although there may still be great variation of detail. More specifically, and a point that we will return to, is that this will also tell us something about what might be involved in modifying the resulting application in the future.

For software the sort of label we use may relate to the way that the application is intended to operate. So we might refer to styles using terms such as 'client-server' or 'pipeline'. However, such labels are not a very systematic way of classifying architectural styles, although very useful for easy reference. And as it became realised that the idea of an application's architecture was useful, so ideas emerged about how best to describe and classify this.

In Perry & Wolf (1992), the authors proposed a classification scheme based on the following characteristics:

software architecture $\leftrightarrow$ *{elements, form, rationale}*

They categorised *elements* as being concerned with processing, data or connections, and *form* as a 'set of weighted properties and relationships', while *rationale* recorded the motivation for particular choices.

The influential book by Shaw & Garlan (1996), that helped to consolidate the idea of architecture and its importance, adopted a rather simpler approach to classification, employing a basic framework based solely upon *components* and *connectors*.

A rather more detailed scheme was used in (Shaw & Clements 1997). This categorised architectural style in terms of the following major features.

- the kinds of components and connectors used in the style (for example, objects and method calls);
- the ways in which control (of execution) is shared, allocated and transferred among the components;
- how data is communicated through the system;
- how data and control interact;
- the type of (design) reasoning that is compatible with the style.

The features described in the last four bullets can be considered as forming the 'context' in which an application will execute, and so we can then categorise architectural styles in terms of:

software architectural style $\leftrightarrow$ *{components, connectors, context}*

It is also worth noting that while researchers have sought to formalise the description of architectural style using 'architecture description languages' (ADLs), this does not seem to have led to anything that is widely used (Ozkaya

2017). Indeed, architectural concepts tend to be rather abstract as well as very diverse, and hence are probably not really easily described using formal notations.

As indicated earlier, applications created in the 1970s and 1980s were usually organised in a self-contained 'monolithic' form, with all of the elements residing on the same computer, and having a form such as call-and-return or data-centred-repository. With the advent of the internet in the 1990s new forms evolved (such as client-server), exploiting the ability to distribute the elements of an application. The later evolution of the cloud, and related ideas about services has added further to the choices available to the designer.

Because architectural concepts are relatively abstract, the vocabulary associated with them lacks precision—one of the factors motivating the categorisation process provided in early works such as (Shaw & Garlan 1996).

Table 6.1 summarises the major categories of architectural style identified in (Shaw & Clements 1997). Perhaps not surprisingly, these largely relate to 'monolithic' forms of application construction—distributed forms can be hard to classify—and many categories do encompass a variety of styles.

TABLE 6.1: Some major categories of software architectural style

Category	Characteristics	Examples of Styles
Data-flow	Motion of data, with no 'up-stream content control' by the recipient	batch sequential; pipe-and-filter
Call-and-return	Order of computation with a single thread of control	main program/sub-programs; 'classical' objects
Interacting-processes	Communication among independent, concurrent processes	communicating processes; distributed objects
Data-centred repository	Complex central data store	transactional databases; client-server; blackboard
Data-sharing	Direct sharing of data among components	hypertext; lightweight threads

One caveat emphasised by Shaw & Clements (1997) and one that we should note, is that applications are not necessarily 'pure' in terms of their style and so may possess the characteristics of more than one style. And of course, if a system is classified as being a 'hybrid', that doesn't reduce the value of the classification process, indeed it may even render it more important.

Bass, Clements & Kazman (2013) describe how thinking about architecture has subsequently evolved. Many of the original ideas are still there, but they are apt to be expressed differently. Rather than categorising styles in terms of their elements and the connections between them, they argue that architecture should be viewed in terms of three different *structures*, which in turn relate to different sets of decisions that occur during the process of design.

1. *Module* structures consist of the blocks of code or data that will form the application, and relate to design decisions about how responsibility is allocated, the ways that the modules depend upon each other (the forms of coupling), and the use of external modules.

2. *Component-and-connector* structures are more concerned with how the application will behave when it executes and how the different elements will interact during execution. Whereas module structures are essentially static, these are concerned with system execution.

3. *Allocation* structures are concerned with where different modules will reside, and from where external resources will be sourced.

While this can be simplified to a model of 'component-connector-context' it goes much further and relates much more closely to the decision-making processes involved in architectural design.

To help illustrate the concepts involved, we now look briefly at the forms of three widely used architectural styles. Other styles such as object-oriented and service-oriented architecture will be discussed in later chapters, and these examples have been chosen simply to illustrate the variety that can occur.

6.2.1 Pipe-and-filter architectural style

An application based upon this style is one in which a sequence of transformations are performed upon data, typically by a series of processes. Each process performs its allocated task and then passes the data on to the next one. As a style it emphasises *separation of concerns*, where the *concerns* are provided by the transformations embodied in the individual processes. It is organised around *flow of data*. Table 6.2 summarises the main characteristics of the style using the classification scheme from (Shaw & Clements 1997).

TABLE 6.2: The pipe-and-filter architectural style

Feature	Instantiation in pipe-and-filter
Components	Data transformation processes.
Connectors	Data transfer mechanisms (e.g. Unix pipes, files etc.).
Control of execution	Typically asynchronous. Control is transferred by the arrival of data at the input to a process.
Data communication	Data is generally passed with control.
Control/data interaction	Control and data generally share the same topology and control is achieved through the activity of transferring data.
Design reasoning	Tends to employ a bottom-up approach based upon *function* due to the emphasis placed upon the filters (components). Fits well with incremental development forms.

Figure 6.1 illustrates the use of the Unix 'pipe' mechanism, introduced in the early 1970s, to provide an example of an application that uses this style. The pipes provide data streams linking the 'standard output' and 'standard input' of successive processes (the 'filters'), labelled as P1, P2 and P3 in the figure. An example of the type of application well-suited to this style is a simple compiler, in which a succession of operations gradually transforms the data stream from input text to output code. So here, P1 might be a lexical analyser, used to separate the 'tokens' in the input program; P2 could be a syntactic checker, ensuring that the tokens obeyed the syntax rules for the language; and P3 might perform a semantic analysis of the stream of tokens, leading on to further processes to generate and optimise code.

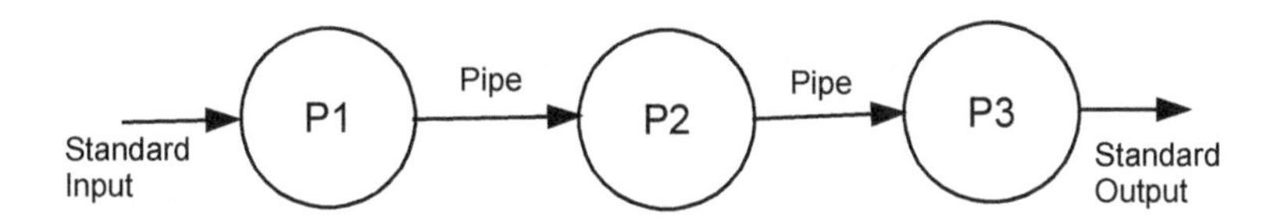

FIGURE 6.1: Unix pipe and filter using three processes: P1 | P2 | P3

As an architectural style this is more versatile than the example might imply and is well suited for use with data-driven operations. Data transfer does not necessarily have to be asynchronous, and the topology is not confined to linear sequences of processes. It also provides scope for strong separation of concerns as far as processing is concerned, since each process usually occupies its own unique address space with no shared data. (However, knowledge about the form(s) of the data does need to be shared.) There is also ready scope for reuse of processes, because of the strong decoupling. It is also well suited to a development approach based on prototyping, and offering scope to perform this incrementally, with each process being developed and integrated in turn.

6.2.2 Call-and-return architectural style

This architectural style is one that closely mirrors the way that the underlying computer operates, using an ordered and hierarchical transfer of control between different processing tasks. Again it is based upon a philosophy of *separation of concerns*, but unlike pipe-and-filter, the main emphasis lies upon the transfer of *control* between processing activities, rather than transfer of data, and so the 'concerns' being partitioned are essentially those related to function. Imperative programming languages (such as Pascal, C. and many others) provide structures that implement this form, whereby a 'main' program element transfers control to a network of sub-programs. Typically, sub-programs either access data through the passing of parameters (arguments) or through the use of static global variables that are visible to all sub-programs. Object-oriented languages such as Java often use this as an underpinning mechanism, and so can also be used to write software in this style.

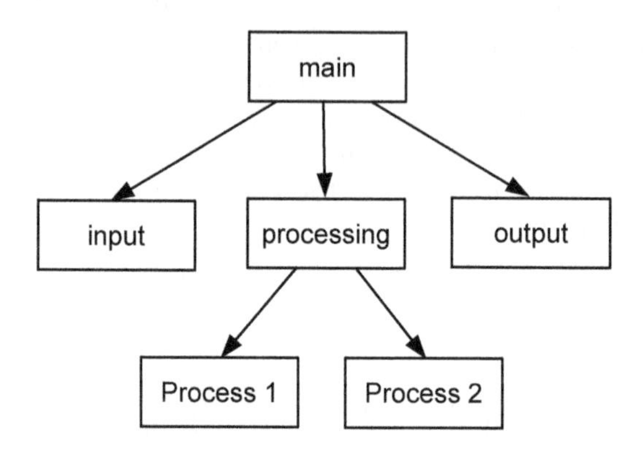

FIGURE 6.2: Call and return using a main program and sub-programs

Figure 6.2 shows a simple illustration of an application structured in this style, with a main program unit (which sequences the calls to the sub-programs), and a hierarchical set of sub-programs that perform the tasks. Here the arrows indicate transfer of control by a conventional sub-program invocation mechanism, and further annotation can be used to show transfer of data via sub-program parameters.

Table 6.3 summarises the main characteristics of this architectural style. While this is a simple and convenient form to employ for constructing monolithic applications, and likely to produce efficient code because of the way that it 'mirrors' the underlying computer mechanisms, it provides limited scope for reuse beyond employing a 'library' of sub-programs. However, it is quite versatile and widely employed—for example, it could be used to structure the internal forms of the processes used in Figure 6.1.

6.2.3 Data-centred repository architectural style

This style encompasses those applications for which the key element is some central mechanism that is used for the persistent storage of information (not necessarily a database management system), so that the information can then be manipulated by a set of external processing units. So for this architectural style, the 'concerns' are related to data and its use. However, it also offers scope to provide an element of *information hiding* regarding the way that the information is actually stored.

Obvious examples of applications in this category are database systems and blackboard-style expert systems (here the blackboard itself provides the central mechanism). Shaw & Clements (1997) argue that client-server systems also come into this general category, on the basis that the server acts as a

TABLE 6.3: The call-and-return architectural style

Feature	Instantiation in call-and-return
Components	Sub-program units.
Connectors	Invocation of sub-programs.
Control of execution	Sequencing of calls is organised by the algorithms in the components that form the invocation hierarchy.
Data communication	Data can be passed through parameters and may also be shared using a global storage mechanism (variables declared in the main component).
Control/data interaction	This is largely limited to the passing of parameters and the return of data through function calls.
Design reasoning	Because the 'concerns' (functions) are sub-divided, it is usually associated with top-down thinking. Many traditional plan-driven design approaches produce design models that have this form. Use of global data structures can constrain the effective separation of concerns.

central repository, and may well include a database. Figure 6.3 provides an illustration of the general form of a client-server system (whether the server is 'local' or in the cloud is not important, since it is the role that matters).

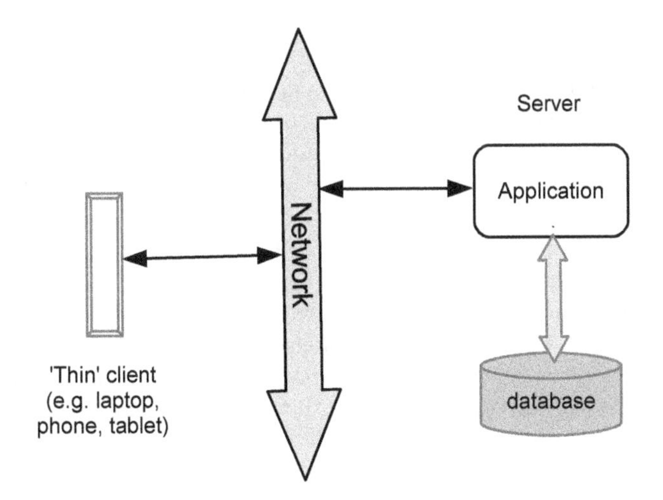

FIGURE 6.3: Data-centred repository: a simple client-server model

When the repository is an actual database, the central mechanism is likely to be concerned with providing one or more forms of indexed access to a potentially large volume of data. Blackboard systems and object-oriented databases can provide a less highly structured form of repository, which may well be used to acquire, aggregate, and record a variety of forms of 'knowledge'. Table 6.4 summarises the characteristics of this architectural style.

TABLE 6.4: The data-centred repository architectural style

Feature	Instantiation in data-centred repository
Components	Storage mechanisms and processing units.
Connectors	Transactions, queries, direct access (blackboard).
Control of execution	Operations are usually asynchronous and may also occur in parallel.
Data communication	Data is usually passed through some form of parameter mechanism.
Control/data interaction	This varies quite widely. For a database or a client-server system, these may be highly synchronised, whereas in the case of blackboard there may be little or no interaction.
Design reasoning	The strong focus on data usually means that modelling the data is an important element. The variety of forms such systems take tends to mean that design logic may also vary extensively.

An architectural style for the CCC

The CCC system is essentially one that is used to manage information—about the cars, the members and the way that members are charged for the use of the cars. So the choice of a *data-centred repository* style seems to be appropriate, providing a central 'knowledge base' that can be accessed to manage bookings, accounts etc. And since the customers will be using remote forms of access such as mobile phones, it also seems appropriate to adopt a *client-server* form for this. One question that this then raises is "what form will the server take?". We could adopt the use of a relational database to hold details of cars, bookings etc., but equally we could use an object model in which we have objects that represent cars. The latter is likely to prove more flexible. We can see this could cope with different models of car using inheritance if CCC later decide to offer different options, and an object model does encapsulate all of the relevant data about a car within an object. However, given the various things that the CCC system does (and may do in the future) it may well be that some form of compromise would be a good way to handle this. The more changeable activities related to cars could be modelled as objects, but things like recording bookings and charging for use might employ a database.
At this stage we will focus on the choice of style and fill out the details later, deferring decisions about these until we have a clearer picture of how the CCC system will work and the directions in which it will later expand.

6.3 Architectural patterns

The concept of using some form of abstract 'pattern' to transfer knowledge about useful design models that could be used to 'solve' issues that are repeatedly encountered across a range of applications emerged in the 1990s, and has been realised in a number of ways. It has been widely adopted for use in object-oriented software development, as a means of describing solutions to recurring challenges that might face the designer (Gamma, Helm, Johnson & Vlissides 1995), usually involving particular parts of a system or application. It has also been suggested that patterns could be adapted to perform a similar role in the design of service-oriented applications (Erl 2009). We will discuss the pattern concept in much more detail in Chapter 15, but in this section we discuss how the concept can be used to describe the overall form of a system, by making use of *architectural design patterns*, a concept proposed by Buschmann, Meunier, Rohnert, Sommerlad & Stal (1996).

The idea of using patterns to help with transferring knowledge about design 'solutions' is rooted in the work of the architect Christopher Alexander (Alexander, Ishikawa, Silverstein, Jacobson, Fiksdahl-King & Angel 1977), and has been described by him in the following words.

> *"Each pattern describes a problem which occurs over and over again in our environment, and then describes the core of the solution to that problem, in such a way that you can use this solution a million times over, without ever doing it the same way twice."*

Essentially, a pattern provides a generic description of, and 'solution' to, a problem, and being generic, it will always need to be instantiated in a way that addresses the specific situation that the designer is actually facing. (This fits in with the concept of 'design as adaptation' introduced in Chapter 1 and Figure 1.2.)

Couched in terms of knowledge transfer, a pattern identifies a particular design problem that might be encountered, together with a strategy for addressing that problem. We also expect that the description of a pattern will identify some possible consequences of adopting the pattern, since the choice of whether or not to use a pattern may well involve making trade-offs between different aspects of the design model.

In some ways using patterns has similarities to the sort of learning involved in the traditional master/apprentice relationship. Learning about patterns involves learning about possible problems that a designer may face as well as about how these might be resolved—usually involving relatively manageable 'chunks' of knowledge. Patterns also have to be *learned* (an issue that we return to when considering the use of object-oriented design patterns). The idea of a pattern also fits well with some of the observed practices of designers, such as the notion of 'labels for plans' that was observed by Adelson & Soloway

(1985), whereby a designer would note the presence of a sub-problem for which they already had a solution strategy, (again, reflecting the idea of designing by adaptation).

At the architectural level of software design the pattern concept is used slightly differently. Here, patterns offer models for the general structure that the application should take, rather than for a part of the design solution. The ideas involved in architectural patterns do overlap with architectural styles to some degree—some patterns can be realised using a variety of styles, whereas others are implicitly linked to the idea of using a particular type of design element (the most obvious example of such an overlap is that of 'pipes-and-filters' which is essentially a pattern that can be employed with processes).

Architectural patterns provide different ways of 'separating the concerns' for an application, aiming to organise its structure so that future changes can be made with a minimum impact from any side-effects. However, to do so successfully does implicitly assume that the designer (system architect) is able to envisage the ways in which the application may later evolve—and a possible risk is that the architectural pattern adopted may not be well suited to what actually does happen. So, while patterns do offer benefits, the adoption of any architectural pattern, like the adoption of an architectural style, does create the potential for generating technical debt.

Table 6.5 summarises the roles provided by a selection of the patterns introduced in (Buschmann et al. 1996). In the rest of this section we examine two examples from this set in order to illustrate the idea more fully.

TABLE 6.5: Some examples of architectural patterns

Pattern	Description of Role
Model-view-controller	Provides a structure for decoupling the various elements of an interactive system.
Layers	Used where an application can be organised as a hierarchy of sub-tasks performed at different levels of abstraction.
Pipes-and-filters	For systems that process a stream of data (this is rather similar to the architectural style).
Broker	Used to structure the interaction of distributed systems with highly decoupled components, where the choice of components may be made 'on the fly' when the application executes.
Reflection	Intended to support dynamic adaptation of the structure for a system.

6.3.1 Model-view-controller (MVC)

This is a widely encountered pattern used to structure the sort of applications in which end-users will interact with some form of 'data repository', performing updates to the data and viewing different projections from it. As an example of where it could be used, it is well suited for use with applications which involve extensive user interaction with a dataset, such as a spreadsheet or word processor.

This pattern is motivated by the principle of separation of concerns, and is concerned with keeping the details of the *model* (the data repository) independent from the elements that need to interact with it, namely the *views* and *controllers*. Essentially the model is encapsulated in such a way that its details (and format) are not directly visible to the views and controllers (the link to designing with objects may be fairly obvious here, since both place emphasis upon *information hiding* through the encapsulation of data).

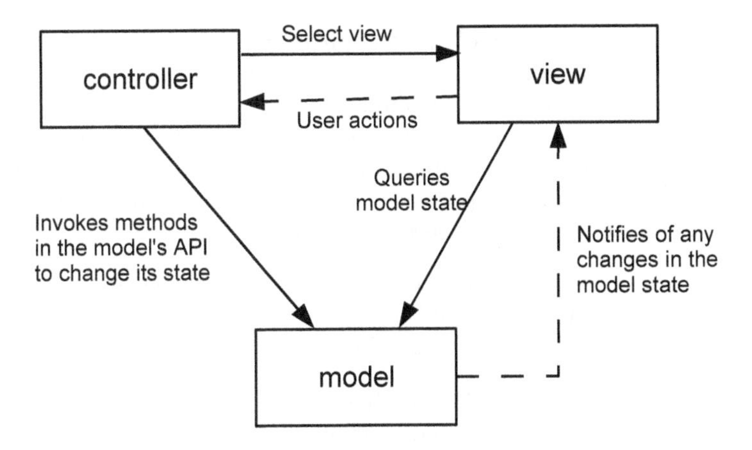

FIGURE 6.4: The MVC pattern

Figure 6.4 shows the idea of MVC in a rather abstract way. It is worth noting that a wide variety of such diagrams exist, often providing slightly different interpretations of the elements making up MVC and how they interact. The roles for the elements are as follows.

Model. This provides the core functionality of the application (that is, the operations performed upon the data) and the associated data.

View. Each view provides a different way of presenting knowledge about the model, obtained from the model, to the user.

Controller A controller handles user input, and each view is likely to have a distinct controller associated with it.

The views and controllers make up the user interface and are largely independent of the model, beyond sharing knowledge of the data types making up the model. The model may well also need to provide updates to the views and controllers when changes are made to it (or at least, provide them with information that an update is available). An important characteristic of this design structure is that it makes it possible to provide quite different 'look and feel' forms of interaction without a need to modify the model itself.

Using MVC for the CCC system

We can easily see that the software for the CCC fits well with the MVC pattern. We can interpret the elements as follows:

- The *model* is the information held centrally about the cars, bookings, customer accounts etc. This is continually being modified and updated as cars are claimed, customers make bookings, cars are taken out of use for servicing etc.
- There are a number of different *views*. Some are ones presented to a customer, perhaps reporting which cars are nearby and available, or present data pertaining to a recently completed hire session. Other views are available to the managers, accounts department, maintenance team etc. They all relate to the same core set of information, but present different elements of it in ways that are meaningful to the needs of a particular user.
- The *controllers* relate to interaction with different users who are making requests, choices etc. Some may be phone apps, others may be devices in the 'back office' of CCC.

An important feature of MVC is that the characteristics of the model as seen by different users through different views should be consistent. This alone is an excellent reason for thinking of the model as a cohesive whole.

6.3.2 Layers

Opportunities to employ the *Layers* pattern occur less commonly than those for using MVC. However, as an example, it illustrates a different way of thinking about how to organise the separation of concerns within the structure of an application, placing greater emphasis upon function.

This pattern structures an application by separating its functionality into different groups of operations, with each group performing its tasks at a different level of abstraction within a hierarchy. A classic illustration of this is the way that networking software is usually organised. Figure 6.5 shows a simple illustration of this for the OSI model, highlighting how each layer provides more abstract operations by performing its tasks by using operations from the layer beneath it.

As an example, when Layer 4 (Transport) splits a message into a series of packets, and ensures that these are delivered to the recipient, it does so by drawing upon the services of Layer 3 (Network), to organise the route(s) by which the packets are delivered.

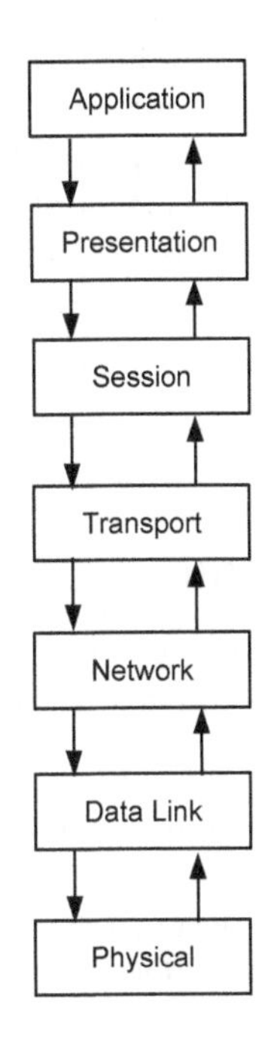

FIGURE 6.5: Layers pattern

The use of *Layers* makes it easy to isolate the effects that can arise from making changes at the lower levels. A simple example concerns a change in the underlying network form. To move from using (say) an ethernet implementation to using a slotted ring as the network medium only involves changes to the very lowest layers. The higher layers have no knowledge of the nature of the actual transport mechanism being used.

Many operating systems use a similar model (for example, Unix applications sit above the 'shell', which in turn makes use of the system layers below it). Again, a different interface can be provided by simply changing the shell, and hence a single computer can readily have a situation where different users are working with different interfaces, but are still able to share files and other resources when necessary.

While clearly *Layers* is a particularly useful pattern to use for 'system' software, in principle it is by no means restricted to such applications. However, it is likely that most designers will encounter this as a pattern that is already established for a software system, and their concern will be to develop applications that conform to it.

6.4 Empirical knowledge about architecture

The nature of software architecture makes it difficult to perform empirical studies, since for each application, the choice of a suitable architecture is rather entwined with the unique nature of the application itself. Hence systematic reviews in this area are relatively uncommon, and those that are available, mainly look at the 'robustness' of architectures when changes need to be made.

An example of this is the systematic review by Williams & Carver (2010) that focused upon assessing the impact of proposed changes to a system's architecture. A further example is provided by Breivold, Crnkovic & Larsson

(2012), that looks at the effect of architecture upon software *evolution*, and at the different approaches that have been proposed to help address this.

A somewhat different and interesting perspective is provided in the systematic review by Shahin, Liang & Babar (2014). This examines some of the forms used for *visualising* software architecture, as well as different elements (such as patterns), and the ways that these are used. They particularly observed that these were often used to help with different aspects of recovery of architectural knowledge about existing applications as well as to help with evolution.

Key take-home points about software architecture

The concept of *software architecture* offers a high-level perspective upon the way that a software application is structured and organised. Some key issues related to this include the following.

Separation of concerns. The major elements of a system will address different 'concerns' and choice of a suitable architecture will reduce the likelihood of side-effects arising from interactions between the elements as well as providing a way to reduce future technical debt.

Composition. The architecture of a system is concerned with how this is to be *composed* from a (possibly diverse) set of elements.

Vocabulary. Architectural concepts provide a useful *vocabulary* that can be used to aid understanding and discussion related to the high-level structure of an application. This may also help identify how it might evolve in the future as well as what scope there is to reuse elements.

Categorisation. A useful way to think of an architectural style is in terms of its *components*, the *connectors* that map the dependencies between the components, and the *context* that describes how the elements execute and relate to each other.

Architectural patterns. These describe some commonly used ways to organise the top-level form of an application.

Empirical knowledge. Such knowledge is mainly concerned with the interplay between the choice of an architecture and the ease of making changes to a system.

Chapter 7

Modelling Software Properties

In the previous chapter we considered how a software application might be structured, using very abstract levels of description. This is because, at the architectural level, the design models we create are largely concerned with describing how an application is to be organised as a set of major components.

In this chapter we look at how the attributes of software can be modelled at a more detailed level, although it is still very much an exercise in design abstraction. In particular, we concern ourselves with identifying the different characteristics of software that we want to describe, because at this level of abstraction we need to think about many different aspects of software.

We begin by examining the issue of what a *model* is, then consider some of the practicalities involved in developing a model. The rest of this chapter is then largely concerned with *what* properties and attributes we need to model and some ideas about how these can be described.

7.1 What is a design model?

In Chapter 2 we introduced the idea of the *design model* as some form of abstract description for our ideas about how to 'solve' some sort of (ill-structured) problem. The idea of using a model of the intended product to explore (and record) design ideas is an old one, and was probably necessitated as the complexity of the final product increased. Pharaoh's pyramid designers may not necessarily have needed to produce models, but certainly by the seventeenth century AD the form of such artifacts as sailing ships had become sufficiently complex to warrant the production of quite elaborate and detailed models that could be used to indicate what the shipwrights were expected to create.

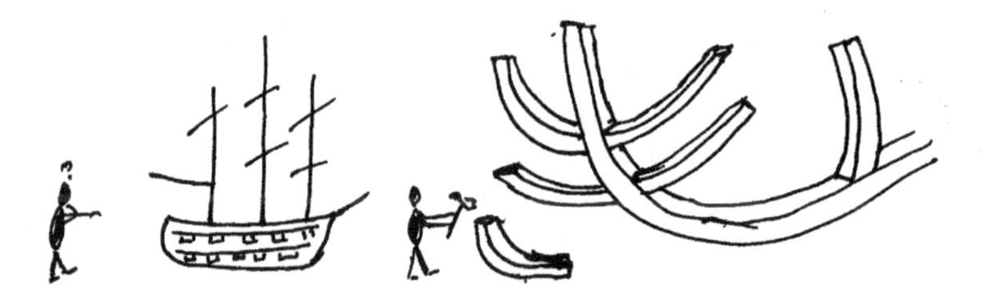

Using a model to guide construction

Software development makes use of a number of different forms of model, not just design models. Modelling is often used to help explore and clarify the *requirements* for an application. Requirements models are usually very abstract, related to the problem rather than the solution, and in some ways play the same role as the models that architects produce to show how their buildings will look and how they will fit into the landscape. We might also use models to help with such tasks as assessing *risk* in the development process, and of course, it is often necessary to use *cost models* to help with planning and managing actual development.

All of these are inter-related of course. A requirements model sets out much of the context and goals for a design model; and the design model provides some of the context required for any modelling of risk or cost. However, for this chapter we will focus on design models—while recognising that they influence, and are influenced by, other models.

Models of physical artifacts tend to have two (related) characteristics. One is that they are likely to be created on a reduced *scale*, the other is the use of *abstraction*. A model is rarely complete in every detail. Usually it possesses just enough detail to show others what the designer intends, but omits issues that are not necessary, perhaps because they are sufficiently 'standard' that those responsible for construction/implementation will understand how they will be provided in the final product.

Modelling is where software once again diverges from the norm. To start with of course, software is invisible, making it difficult to relate our models to the final product in the way that we can look at (say) a scale model of a sailing ship and envisage what the actual ship will be like. And for the same reason, there tends to be no real sense of scale when we create software models. And because software is dynamic, we also have to capture our ideas about a number of quite different attributes, making it difficult to have a single model that will show our ideas. Certainly though, abstraction is something that is important in software models—an important element in the task of modelling software is to leave out the detail that we don't need in order to understand what an application is to do and how it will do it.

Of course, design engineers in other disciplines don't just concern themselves with physical models, and indeed, mathematical models of such issues as stress and dynamic behaviour provide important abstractions for them. However, even then, their mathematical models will be related to the physical properties of the intended system, as well as upon the laws of nature.

This is therefore a good point to clarify some of our terminology. The concept of something exhibiting a particular *property* is quite important in modelling (and also in measurement). We also make use of the term *attribute* in much the same sense, and indeed, the two terms are often considered to be synonymous in everyday language. For our purposes though, it may often be useful to make a distinction between the two terms, rather akin to that sometimes used in the context of measurement. So in this book we will use these terms as follows.

Property. A general concept relating to some characteristic of software (actual or desired). For example, we may describe an application as having the property of being *robust*, or *efficient*.

Attribute. This is something that we can explicitly and directly model (and measure). For example, the average time to respond to an event, or the form of coupling occurring between two design elements.

In the context of modelling, this allows us to distinguish between what we can and cannot model directly. Design models largely describe or make reference to attributes—whereas a model produced as part of a specification would probably largely refer to properties. So the distinction is useful in terms of clarifying what we can and cannot readily model.

So, how do we model software and its attributes (or properties)? Well, this will be explored in detail in the following sections and the later chapters, but we commonly use three forms of notation (often in combination).

- Software designers make extensive use of *diagrammatical* notations. Often these are informal 'box and line' sketches used to help explore ideas. At other times we draw rather more stylised diagrams to help explain ideas to others. In this chapter we explore ideas about this type of notation to describe ideas about modelling.

- More formal modelling sometimes makes use of *mathematical* notations to describe system properties and behaviour. While such models are less well-suited than diagrams to exploring ideas, they are much better at supporting reasoning about the consequences of design choices.

- Narrative descriptions using *textual* forms may also be useful, and while they can be used in their own right, these are almost always needed to supplement the other two forms.

We can use these models for all sorts of purposes, including exploring ideas; explaining them to others; and checking them for completeness and consistency. And of course, as always with design, there are no 'right' and 'wrong' choices of notational form in any absolute sense.

7.2 Representations, perspectives and viewpoints

Abstraction performs an essential role in the design process, allowing the designer to concentrate on those features of the design model that are most important at any particular stage of its development. The designer therefore needs ways to represent abstract ideas about problem objects and design objects, and about the different relationships that will occur between them. While this is true in any design context, the invisible nature of software poses some additional challenges, as observed earlier.

When modelling, a *representation* is used to provide a particular abstract description of particular characteristics of the designer's ideas for an application, and is typically needed for such purposes as:

- allowing the designer to express their ideas about their 'solution';
- helping the designer to explain their ideas to others (such as customers, fellow designers, implementors, managers);
- assisting the designer to check the design model for important features such as completeness and consistency.

A representation, whatever its form, can help a designer to concentrate his or her attention upon those issues that are most important to them as the model evolves. It also provides a mechanism for reducing the extent of the 'cognitive load' involved in the process. Indeed, as Vessey & Conger (1994) have observed: "human problem-solvers resort to various ways of effectively increasing the capacity of short-term memory as well as using external memory devices and/or techniques to support working memory". In this context, a representation can be considered as a form of 'external memory device'.

Two useful and related concepts are those of *perspectives* and *viewpoints*.

- A *perspective* relates to a software development role, such as a manager, end-user, programmer. Each has their own set of interests and needs related to the design model.

- A *viewpoint* relates to a set of particular characteristics of a design model, which in turn reflect some specific aspect of software (its construction, the way it behaves when certain events occur etc.). We usually use specific representations (notations) to describe the design model from particular viewpoints.

Our concern in this chapter is primarily with viewpoints, but it is worth pointing out here that the concepts of perspectives and viewpoints are not wholly independent. A specific perspective of the design model can be presented using a particular set of viewpoints and representations.

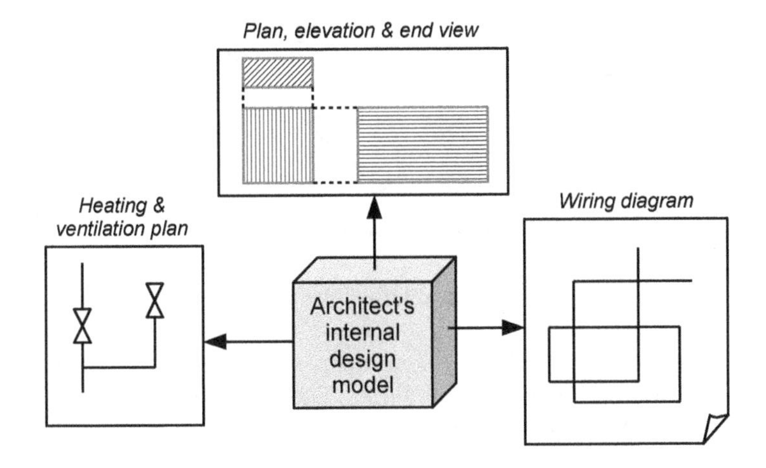

FIGURE 7.1: Examples of representations as realisations of viewpoints

In this chapter these terms are also used rather differently from the way that they are used in requirements engineering, where a viewpoint may be used to describe a 'stakeholder' view of the design model (Finkelstein, Kramer, Nuseibeh, Finkelstein & Goedicke 1992). In this book the concept of a viewpoint is more focused upon the design model, and with providing *projections* of its properties outwards to users, whereas its use in the requirements context is more concerned with a user's particular *perception* of the model. Arguably this stems from the use of 'white box' role of models in software design, where the role of the model is to describe the workings of the application, whereas the model in requirements engineering is more of a 'black box', whereby we describe *what* a system is to do, rather than *how* it is to be done.

If we very briefly return to the example of designing a house move, used in Chapter 2, examples of design viewpoints appropriate for this may include its physical layout, the organisation of electrical power, and heating and ventilation. These are all different projections of the 'house model', as shown in Figure 7.1. Since each representation used to model a viewpoint has some overlap and dependency upon the other forms, (for example, heating and ventilation relates to the physical layout of the interior of the building) these can really be seen as forming projections of the model, and even for this physical situation, there is a clear need to ensure consistency between them.

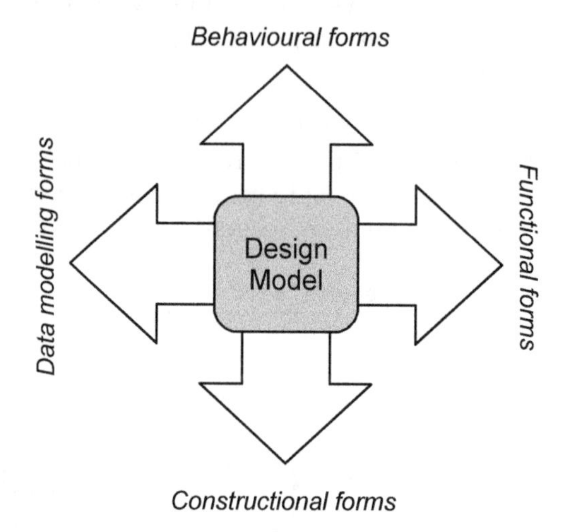

FIGURE 7.2: Examples of representations as realisations of viewpoints

In Figure 7.2 we identify the four main viewpoints that we will use to categorise design notations in this book. The attributes that each of these models can be described as follows.

- *Constructional* forms are concerned with how the elements of the model are composed together to create the application, and hence model the different ways in which the elements are coupled and the resulting dependencies. This is essentially a *static* projection from the design model.
- *Behavioural* forms describe the causal links that connect events (internal and external) and system responses. As a viewpoint it models the *dynamic* interactions arising from events that cause transitions between the different states of a system.
- *Functional* forms describe the actions performed by a system or application. These may depend upon such aspects as information flow, or the sequencing of actions, depending upon the nature of the system and its architectural style.
- *Data-modelling* forms are concerned with the different data elements used in the system and the relationships that occur between them.

What we need to remember is that this is a classification scheme rather than some fundamental characteristic of software. What the different viewpoints 'capture' is some of the different characteristics of software, and hence of our ideas about the design model itself. And since each representation used to describe a viewpoint describes a particular set of design attributes, there may be some overlap between these, as shown symbolically in Figure 7.3, providing some potential for checking consistency between viewpoints with respect to the model itself.

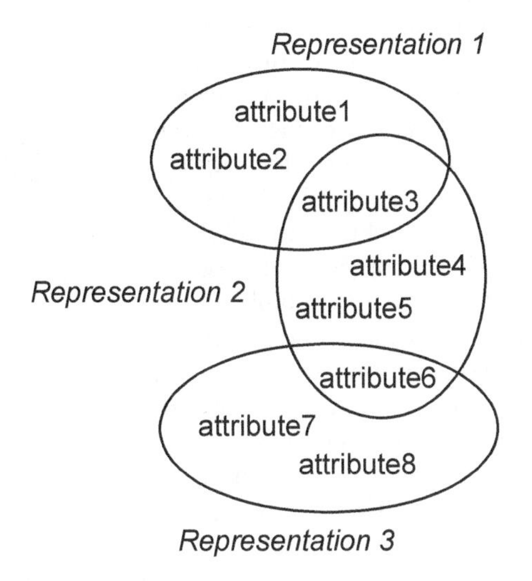

FIGURE 7.3: Some attributes may appear in more than one representation

What is more, these viewpoints can themselves be further classified as being 'direct' viewpoints, in that they are created directly by the designer(s). We can also identify viewpoints that may be described as being 'derived' viewpoints, created by applying some form of transformation to the design model. The most obvious such form occurs when some form of interpreter is used to 'execute' the design model (usually a behavioural description of the model), rather akin to the process of *mental execution* noted earlier and described in (Adelson & Soloway 1985).

While this scheme of classification is quite practical for our purposes, it is by no means the only way that we can classify design descriptions. A widely-cited form that is used for describing object-oriented models is Kruchten's *4+1 View* that uses the labels: logical, process, development, physical + scenarios (Kruchten 1994). While less general-purpose than the scheme used here, there are some similarities, including the common recognition that some characteristics are not necessarily uniquely represented in one single viewpoint.

In the rest of this chapter we examine the characteristics of each viewpoint and identify the attributes of design elements that are modelled in these. In the chapters that follow, we will then look at some examples of how these viewpoints are modelled for some of the major architectural styles.

7.2.1 The constructional viewpoint

Constructional descriptions are produced at various stages in the development of the design model. *Plan-driven* approaches used with call-and-return architectures will develop constructional models towards the end of the design process, whereas those used with object-oriented styles as well as more dynamic forms of development such as the *Agile* forms may identify the major constructional elements at relatively early stages. And the choice of architectural style at an early stage of any development process essentially constrains the choices available to the designer. (Plan-driven methods tend to be organised around a specific architectural style, so adopting a particular plan-driven design approach also determines the choice of constructional form.) Constructional information is also dependent upon the particular forms of *packaging* used in an architectural style (classes, sub-programs, processes etc.) and how they interact.

FIGURE 7.4: Construction and construction units

When thinking about how to actually construct software applications, and the way that the different elements interact, our ideas are likely to be concerned with *coupling* and the different forms it can take within the organisation of an application. Table 7.1 identifies some of these forms, as well as what needs to be described for each of them.

TABLE 7.1: Some forms of coupling used in construction

Form of Coupling	Description
Invocation	The key attribute being modelled here is that of *transfer of control*, whether this be between sub-programs (call-and-return), methods, processes, objects, services or any other packaging form related to executable code.
Uses	This is a rather general relationship which might involve various forms of dependency: invocation, obtaining data, using data types or some combination of these. The important aspect is that this is direct *usage* rather than through forms such as inheritance.
Inheritance	Inheritance is usually associated with an object-oriented style and is concerned with reuse and adaptation of some existing attribute.
Data Flow	This requires that two elements (often processes) share knowledge about the form of some data exchange.
Data Sharing	As implied, the coupling involves knowledge of some common data forms or data itself.

Notations used to model this viewpoint may well combine different forms of coupling (such as *uses* and *inheritance*. (These two forms can also be considered as being rather more abstract than the other three.)

More generally, this viewpoint is largely concerned with modelling static structures rather than with any form of run-time behaviour. Even when the form modelled is something like *invocation*, the concern here is *what* is being invoked, not *when* it is invoked. And as a result, when we look at some examples in the following chapters, we will find that these are heavily influenced by architectural style.

7.2.2 The behavioural viewpoint

This viewpoint is concerned with causal relationships, connecting 'events' to 'responses' via any necessary conditions. The associated notations tend to be far more abstract than those used for construction as the events and transitions may involve operations that are spread across a number of different constructional elements.

Many behavioural notations can be considered as examples of *finite-state machines*, being concerned with the transitions that occur between different states of an application (such as *waiting*, *processing*, *generating output* etc.), and the changes in context (events) that are required to make these transitions occur. And being concerned with dynamic relationships, these forms are useful when describing the attributes of a design model that are concerned with time in any way. However, not all aspects of timing can be handled with equal ease:

- aspects concerned with *sequencing* of events can be described fairly well;
- descriptions related to *fixed-interval* events are also fairly tractable;
- the effects of timing *constraints* are difficult to represent.

Behavioural descriptions do not necessarily require that the design model has been elaborated beyond being able to provide a 'black box' description of a system, particularly when thinking about events in the abstract. However, they can also be used for quite detailed modelling of how system elements will be affected by events, both internal and external, and as such, this class of representations is an important one.

7.2.3 The functional viewpoint

One of the harder tasks for a designer is that of describing just *what* an application does. It is definitely more challenging than describing how it is made up of software elements (construction) or how it will behave when some event occurs. This is largely because this viewpoint is closest to the actual purpose of producing the application and hence is strongly related to the 'problem' it is seeking to address (booking cars, providing an auto-pilot for an aircraft, monitoring a patient's blood pressure, etc.) and so involves a strong domain element.

However, this is also something that is likely to be described quite well in the *requirements specification*, although again, how it can be described may be quite challenging. So functional descriptions of a design model are apt to be expressed in terms of algorithms, data manipulation, and other software-related activities.

This also highlights one of the challenges for software designers. One of the viewpoints that everyone would like to be able to model is probably the hardest one to model! So, much of this has to be captured in the other viewpoints, making use of those characteristics where relevant attributes overlap.

7.2.4 The data-modelling viewpoint

While the description of data structures need not be a very significant issue when undertaking architectural design (apart from when we are developing such intrinsically data-centric applications as database management systems of course), it is often an important aspect for detailed design. Important attributes that may need to be part of the design model include *type* (both in compounding data types to create new ones and in such mechanisms as inheritance, used to create new classes), *sequence* (in terms of structures such as trees and lists), and *form*.

For some classes of problem, the choice of data structures is a central one and cannot be easily divorced from the functional aspects. Like construction, data-modelling is largely concerned with *static* relationships, and so can sometimes be modelled using similar or the same notations as those used for construction.

7.3 Design notations

Having briefly discussed *what* we want to be able to describe about a design model, the next question is *how* this might be done? As noted earlier,three forms commonly used for design representations are:

- text
- 'box and line' diagrams
- mathematical expressions

Of course, these are not mutually exclusive, and indeed, neither of the latter forms is likely to be of much use without a supporting textual component.

7.3.1 Textual description forms

Text is widely used as a means of summarising information, both when sketching and also when producing more formal descriptions. Ordered lists, tables, and highlight boxes, as used throughout this book, provide a ready means of presenting information. Two problems with using text on its own to describe aspects of a design are as follows.

- Any structure that is implicit in the information can easily be obscured, unless it can be highlighted using lists, tables or boxes. Sub-headings and indentation can help to show structure, but may be less effective over long blocks of text.
- Natural-language descriptions can easily be ambiguous, and structuring text so as to avoid ambiguity can lead to the use of long and complex structures.

Text is really most effective when used in small blocks and tables. We will use some examples of these in the next chapters, but it is also worth noting that the SSADM (structured systems analysis and design method) approach to plan-driven design is one that makes effective use of text through the use of standard pro-formas to record information about certain design decisions (Longworth 1992). (SSADM was developed to meet the needs of local and national government agencies. Its designers recognised that these were organisations where staff were regularly transferred between projects and departments, and hence that design rationale could easily be lost when this occurred. So in specifying its procedures they took care to include the means of recording these. We discuss SSADM further in Chapter 13.)

7.3.2 Box and line description forms

We have already seen various examples of box and line forms, and there will be many more in the chapters that follow this one. As with text, diagrams do benefit from simplicity of form, and diagrams that contain a large number of symbols may be difficult to comprehend. Box and line diagrams are easily sketched on a whiteboard or on paper, but of course, because of the invisibility of software they lack any ready visual link between their form and their meaning. Curiously, despite a long history of using such forms for describing software, there would appear to be little or no research into their form and use.

Many widely used forms have a relatively small number of symbols, possibly reflecting the issues underlying the 'magical number seven plus or minus two' discussed in Chapter 3. More abstract notations tend to have fewer symbols too.

For most users, diagrams are probably the most effective way of providing a clear summary of ideas about the abstract concepts and relationships involved in developing a design model. However, we should remember that this is not guaranteed, and a poor diagram is no more helpful than a block of unstructured and ungrammatical text. Like text, diagrams have both a *syntax* ('how we express an idea') and a *semantics* ('what it means') and these need to be used correctly if ideas are to be conveyed to others. However, this should not be over-emphasised, especially at the sketching phase of model development. Diagrams being used to develop ideas often have very informal syntax and semantics, possibly being created as they are drawn, and these aspects only become more important when we need to record ideas for others.

Indeed, and sometimes frustratingly for a tool-creating culture such as software engineering, this does limit the usefulness of software tools that can be used to create diagrams. Because these tools often tend to enforce syntactic rules (or at least, don't necessarily permit free use of the notation), their most useful role is probably one of record-keeping rather than diagram development.

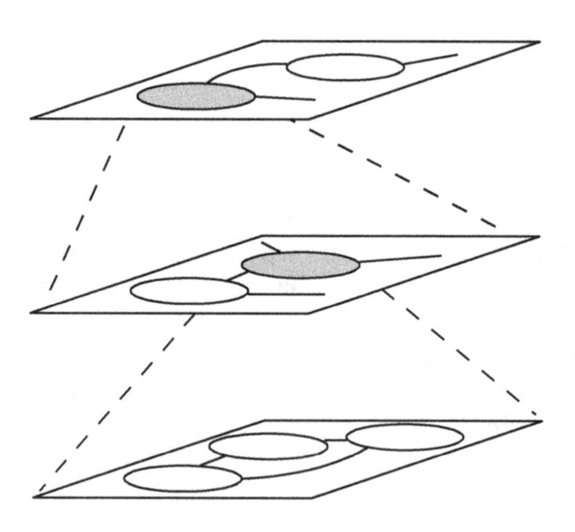

FIGURE 7.5: Hierarchy in diagrams

A useful property of some of the more formal diagrammatical notations is that of a *hierarchical* organisation, shown schematically in Figure 7.5. This occurs where one or more forms of 'diagram element' can themselves be described in an expanded diagram of the same form. A hierarchical organisation offers the advantage that diagrams can be 'layered' into levels of abstraction, avoiding large and complicated diagrams and so aiding comprehension. Again though, this is probably of limited value when sketching, and mainly useful when creating diagrams using tools.

Table 7.2 summarises the details of some well-known diagrammatical notations. Most of them are ones that we will encounter in the following chapters as we look at how to model some widely-used implementation forms.

TABLE 7.2: A selection of box and line notations

Representation	Viewpoints	Characteristics modelled
Data-Flow Diagram (DFD)	Functional	Information flow, dependency of operations on other operations, relations with data stores.
Entity-Relationship Diagram (ERD)	Data modelling	Static relationships between data entities.
State Transition Diagram (STD)	Behavioural	State-machine model of an entity or system. .
Statechart	Behavioural	System-wide state model, including parallelism (orthogonality), hierarchy and abstraction.
Jackson Structure Diagram	Functional, Behavioural, Data Modelling	Forms of sequencing employed for operations, actions and data entities.
Structure Chart	Constructional	Invocation hierarchy between subprograms.
Class Diagram	Constructional	Coupling between classes and objects.
Use Case Diagram	Behavioural, Functional	Interactions between an application and other 'actors'.
Activity Diagram	Behavioural, Functional	Synchronisation and coordination between the actions of an application.
Sequence Diagram	Behavioural	Message-passing between elements and interaction protocols.

7.3.3 Mathematical notations

Mathematics (or at least, mathematical notation) offers scope to combine abstraction with a lack of ambiguity. Various notations for use in specifying the properties of software have been proposed (a good account of the evolution of these forms, and the claims for some of them, is provided in the review by Shapiro (1997)). These *formal description techniques* or FDTs have had some success, particularly where there is a need to ensure the robustness of safety-critical systems, preferably via rigorous proof mechanisms. Probably the best-known formalism is the Z specification language (Spivey 1998). There is a brief discussion of this in Chapter 18.

FDTs have particular strengths in describing system behaviour, as well as in handling some of the issues of time dependency. They can also be supported by tools that help with checking for such things as completeness. However, their use does require some training, as well as some familiarity with discrete mathematics, and so are best described in more specialist texts.

7.4 Empirical knowledge related to viewpoint notations

The rather abstract nature of the material of this chapter means that there are really no directly relevant empirical studies, although there are some studies of specific notations or forms that we will cover in later chapters. However, this is a good point to mention a conceptual tool that can be useful when empirically evaluating the usefulness of notations—*cognitive dimensions*, sometimes also termed *cognitive dimensions of notations* (Green & Petre 1996, Blackwell & Green 2003). This is a set of qualitative measures that can be used to assess different aspects of what is usually termed 'information representation', which of course includes those forms used to provide the description of a design model, with the purpose of the dimensions being to provide a set of 'discussion tools' for evaluation. Here we limit our description to the original set of 14 dimensions described in (Green & Petre 1996) (the systematic review by Hadhrawi, Blackwell & Church (2017) provides a useful summary of some of the key papers describing cognitive dimensions). Table 7.3 identifies the dimensions and provides a summary of their meanings. While these are apt to be couched in terms of HCI (Human Computer Interaction), they clearly also have something useful to offer to design models in general.

Here, we briefly examine some simple examples of particular dimensions, and their application to the process of modelling.

The dimension of *hidden dependencies* is one that can be observed when modelling some forms of architecture, and can reflect a lack of clear encapsulation of design elements. This can be seen when we model 'call-and-return' structures. While our diagrams may indicate which elements are invoked by a particular element, and the details of any parameters (arguments) used to convey information between them, they may omit details of 'global' variables that are also used to share information, and that can affect the way that the model operates.

The wide variety of software architectural styles can make it difficult to avoid some degree of *premature commitment* when designing software applications. The choice of architectural style often needs to be made early in the process, and the need to work with other software applications can also influence the choice of style. This in turn can then affect the order in which we need to make further decisions.

TABLE 7.3: The cognitive dimensions

Dimension	**Interpretation**
Abstraction	Types and availability of abstraction mechanisms
Hidden dependencies	Important links between entities are not visible
Premature commitment	Constraints on the order of doing things
Secondary notation	Extra information provided via means other than formal syntax
Viscosity	Resistance to change
Visibility	Ability to view components easily
Closeness of mapping	Closeness of representation to domain
Consistency	Similar semantics are expressed in similar syntactic forms
Diffuseness	Verbosity of language
Error-proneness	Notation invites mistakes
Hard mental operations	High demand on cognitive resources
Progressive evaluation	Work-to-date can be checked at any time
Provisionality	Degree of commitment to actions or marks
Role-expressiveness	The purpose of a component is readily inferred

The CCC and premature commitment

The concept of *premature commitment* resonates well with the idea of *technical debt.* In Chapter 1 we observed that we might model the information about the cars using a database or using a set of objects. This is just the sort of architectural choice referred to above—and the consequences for other parts of the design model need to be considered carefully before making a final choice.

Arguably some design notations are prone to *diffuseness* (we will see this in the next chapter) in the sense that the notations are overloaded with symbols or notational variations. This puts a greater cognitive demand upon the reader, who needs to remember what a diamond in the middle of a line means, or what the difference is between a continuous line and a dashed one.

This is very much an evaluation framework that can be applied (and has been) to design notations as well as to HCI design.

Key take-home points about design modelling

Modelling involves creating an abstract description that reflects the intended properties of a software application in a form that is visible, and that describes its different attributes.

Design models. Software is usually modelled by using some combination of diagrams, text and formal mathematical notations. Each of these has its particular strengths and limitations, and text in particular is often used to supplement the use of diagrams and mathematical forms.

Viewpoints. These effectively form 'projections' of the design model, describing specific sets of attributes, and are realised through design notations. Constructional, behavioural, functional and data-modelling viewpoints form a useful and practical set of categories for the different attributes.

Cognitive dimensions. These provide a useful set of concepts for evaluating the form of particular notations, and their limitations.

Chapter 8

Sketching Design Models

Having considered how we can model different properties and attributes of software, in the three chapters that follow this one we bring together ideas about architectural style and modelling to consider how the elements of an application can be modelled. However, before doing so, we divert slightly to consider the issue of *sketching*.

The role of this short chapter is to distinguish between the way that many textbooks present the form of a design model, and the way that it is often produced by designers. Modelling can itself be viewed as an ill-structured problem, and in particular there are no right or wrong ways to describe a model. There are definitely times when we need a set of diagrams that conform closely to well-established rules of syntax and semantics, but there are also many times when something informal may quite effectively meet our needs. Both are useful, the question is mainly when to use one or the other.

8.1 Why do designers sketch?

Designers often produce 'rough' (in the sense of informal) sketches of their ideas about design models to help them 'externalise their thoughts' (Petre & van der Hoek 2016). There are many reasons why designers use sketching to rough out their ideas. Sketching helps them with thinking through their ideas, choosing between options, explaining their ideas to others, simulating execution of an application, and generally exploring the problem space, as well as performing various other design-related tasks. So sketches can be used for exploration, reflection and assessment, checking for completeness, and to assist collaboration with others. And sketches allow these things to be done quickly, possibly 'on the spot', without needing to generate formal design documents—indeed, in many ways, the use of sketches complements agile development in avoiding formal document-driven practices.

In this book we are mainly concerned with the use of sketches for describing design models, although some of the other roles, especially reflection and assessment, also play a part in design. We also need to recognise that these different roles are interdependent—for example, deciding between design choices might need some exploration of the problem space to clarify requirements, or the use of mental execution to assess how each option will perform under different conditions.

So, while we do emphasise the use of sketching in this book, our use of it is by no means completely representative of the many ways in which it can be employed during software development. (For more examples of these, see the discussion in the introduction to the study by Mangano, Toza, Petre & van der Hoek (2015).) And in this part of the book, we are mainly concerned with how our sketches can be used to think about the different properties required from the design elements themselves.

8.2 Sketching: developing informal models

In Chapter 4, we noted that one of the characteristics that distinguished expert designers was that they *sketched* (Petre & van der Hoek 2016). And that is why, when introducing the *City Car Club* at the start of the book, we also observed that the illustrations related to the CCC would usually make use of informal sketches, rather than being in the form of neatly produced diagrams. Sketching is what designers do, but despite that, learning to sketch is something that is rarely part of any curriculum.

Sketching is what designers do

So, why do designers sketch? Well, chiefly to ensure that they get the design model right, or at least, as right as they can within the context of their understanding of the problem and its context. This is what matters, not the production of neat diagrams, useful though these might be at a later stage (Petre 2013). (Much the same thing occurs when an author is producing a novel, they focus on the plot and the writing, not on producing a beautiful example of word-processing.) And working with sketches that are easily changed does emphasise that the design process is one based upon the dynamic and opportunistic evolution of ideas about the design model.

One thing that sketching a design model (or part of one) does is to loosen the constraints of syntax. If a designer wants to sketch out an idea based on (say) using a particular set of classes and objects, and chooses to use ovals to represent classes while simply labelling the arcs between these to indicate what form of coupling is being employed, then if it helps the designer work out their ideas, doing this is fine. And it may even be quite acceptable for them to use such diagrams to discuss their ideas with others. In a sketch it usually doesn't matter what shape is being used for a particular item—and the same sketch may be used to represent a variety of attributes.

Sketches may well be accompanied by textual notes (these are often essential) and also by more structured forms of note such as lists, as was observed in the study of expert designers led by van der Hoek & Petre (2014). Extensive use of lists was also noted in a later study by Mangano et al. (2015), where it was noted that lists were often used in preference to diagrammatical notations that could present a more complete picture of the designer's ideas.

However, if our sketches do use an informal syntax, there is a need to ensure that the meaning of it is retained in some way, usually by means of annotation. The value of sketching lies in developing and talking through ideas about a design model, the value of syntax lies in helping us to understand those ideas as expressed through the design model. So if the syntax is informal or in a non-standard form, the form being used does need to be explained in some way. (There is an assumption here that sketches, although they are informal, are not throw-away items. Indeed, since they embody the history of the way that a design has evolved—including any ideas explored and then discarded (and the reasons why)—there is an argument that they should be preserved in some way.)

Another benefit of sketching is that it also helps overcome cognitive limitations. Miller's 'Magical Number Seven' that we discussed in Chapter 3 may not directly relate to this issue, but the idea of helping to manage the cognitive load imposed by developing a design model that it incorporates does. It simply isn't practical to hold a complete design in one's head unless it is very small, but a sketch may well be enough to help the designer recall their intentions about different elements, as well as to be reminded of earlier decisions. And such a sketch may well provide support for 'mental execution' and for checking a design for completeness.

In the rest of this chapter we address the question of what features of a design model should (or might) be sketched. The following chapters then describe some more 'formal' notations, but fairly informally!

8.3 Characterising the design elements

The choice of architectural style plays an important role in determining much of what the designer wants to model. In particular, it determines the forms of 'packaging' used with particular styles, and the mechanisms that the different elements making up an application will use to communicate with each other. So not surprisingly, we find that the way that the constructional and data-modelling viewpoints are organised will be very strongly influenced by architectural style.

When we consider the behavioural viewpoint, this is much less the case, although not entirely so. Behavioural descriptions tend to be rather 'black box', and hence are less concerned with mechanisms than with effects. And to some extent this is also true for the functional viewpoint, where issues related to *what* is to be done are of more concern than *how* it is to be achieved.

And there are other forms of interaction too. Where design elements are loosely coupled, the designer may need to consider how their interactions are to be synchronised. While we will not explicitly address *real time* aspects in this book, for some systems these are of critical importance, although acting largely as an external constraint upon how a model is realised rather than determining the form of the model itself. Again, this is where sketching (along with note-making) may be preferable to complex syntax, at least for forming initial ideas.

8.3.1 Software design as an ISP

Figure 8.1 is a re-interpretation of part of a figure that we encountered in Chapter 1 that described the process of design as an ISP. (To keep the diagram from becoming too cluttered, not all lines have been included for the data modelling viewpoint.) Now that we have an explicit viewpoint model, we can use this to be more specific about the form of the design model used to represent a software application. However, we should also note that there may be other elements of the design included in the model and also that we might use more than one diagrammatical form when modelling a particular viewpoint.

As a simple illustration of this interconnectedness of the model, consider the following scenario from the CCC.

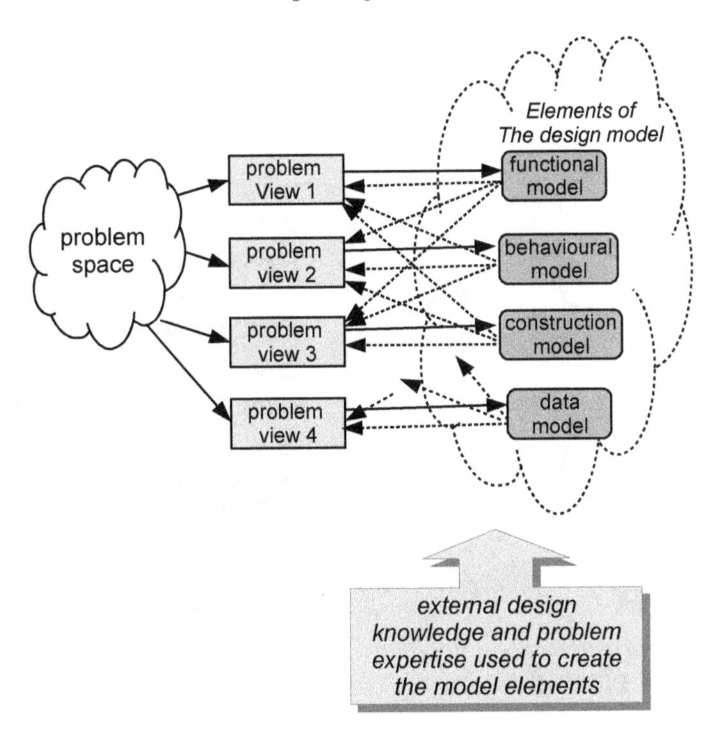

FIGURE 8.1: Software design as an ISP

> **The CCC needs a time-out**
>
> In developing the behavioural model, the designer(s) realise the need to have some form of 'timeout' event, so that if a booked car is not claimed for use within a given period (one that is long enough to allow the user to locate the car and access it), then the booking will be cancelled. This then requires that the functional model be changed to incorporate the setting and clearing of a timer, and that performing these tasks needs to be allocated to some elements of the constructional model.

8.3.2 Sketching initial models

Before looking at some of the different modelling notations and how they are employed, we should return to this chapter's theme of *sketching*. Sketches can often provide a preliminary form of design model, used to clarify ideas as well as to form the basis for more 'formal' descriptions of the different viewpoints. And of course, since such sketches may well never be elaborated more formally, they may also constitute all or part of the design model too.

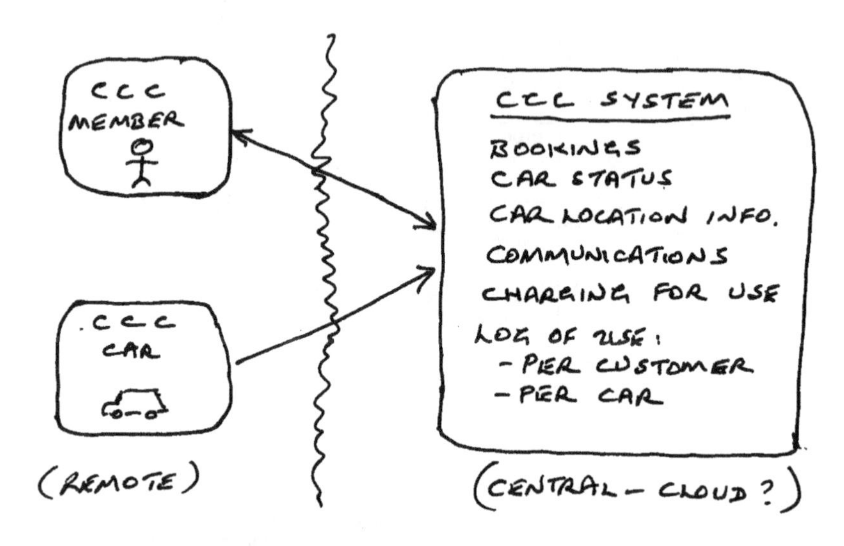

FIGURE 8.2: A very early sketch for the CCC

An important characteristic of many such sketches is that they are 'architecture-neutral'. If we look at the example in Figure 8.2 of the very preliminary sketch for the CCC that we met earlier, we can see that in this very early phase of thinking, ideas are open about what sort of elements the shapes represent from an operational perspective. Beyond the high-level abstractions, they could well be (combinations of) objects, processes, sub-programs, ... What matters is that the design elements are intended to work together to process some form(s) of input and perform such tasks as making bookings.

Further sketches can then be used to create *user stories*, whereby various scenarios of use are used to 'execute' the design model. These stories can address such tasks such as making a reservation, finding a car etc., and may later form the basis for more formal 'story boards' (these are discussed later). Doing so allows the designer to also build up lists of issues that will need to be considered when getting into more detail about the elements, and these lists also form part of the early design model.

Sketches and lists can also be used to help the designer clarify their ideas about some item. Figure 8.3 shows how the designer might use the model in Figure 8.2 to help them think through issues related to the process of booking and using a car. Note particularly that the ideas do not necessarily come in a neat order: having thought through the main steps, the designer then starts to ask questions (such as 'how do we find a car we have booked?').

FIGURE 8.3: Listing ideas about the model

And while this may not appear as architecture-neutral as all that (implicitly, it already looks like a client-server model), it is important to remember that there might be other ways of creating such a system. While it might seem unlikely, the CCC could instal small booths in the streets that users could use to make bookings instead of using mobile phones.

The sketch in Figure 8.4 below summarises some of the things that characterise the use of sketches when used for modelling design solutions. It may be useful to keep these points in mind when reading through the following chapters. As already observed, these forms are very likely to be the eventual outcome from sketches, not something that we start with. And of course, these are 'advice' rather than 'rules' for sketching, since the whole point about sketching is that there are no rules!

The following chapters examine a variety of different modelling forms. We begin by looking at those forms commonly developed for describing *processes*, go on to look at the rather more complex world of *object-oriented* modelling and then take a rather briefer look at how we can model *components* and *software services*. However, the divisions between these are certainly not set in stone. Rather as design sketches can often be architecture-neutral, so we can often use notations in a context that is quite different to the one for which it was developed, as we will see when we look at software services.

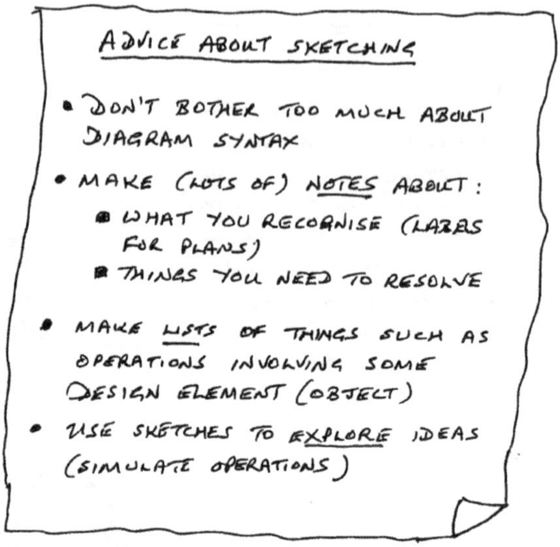

ADVICE ABOUT SKETCHING

- DON'T BOTHER TOO MUCH ABOUT DIAGRAM SYNTAX
- MAKE (LOTS OF) NOTES ABOUT:
  - WHAT YOU RECOGNISE (LABELS FOR PLANS)
  - THINGS YOU NEED TO RESOLVE
- MAKE LISTS OF THINGS SUCH AS OPERATIONS INVOLVING SOME DESIGN ELEMENT (OBJECT)
- USE SKETCHES TO EXPLORE IDEAS (SIMULATE OPERATIONS)

FIGURE 8.4: Advice about sketching

8.4 Empirical knowledge about the use of sketching

The use of sketching is something that is of interest to a number of research communities, particularly those concerned with the intersection between computing and the 'soft skills' used to study human behaviour.

Sketching plays quite an important role in the book on designer behaviour from Petre & van der Hoek (2016). And the more formal study from the same authors (acting as editors) provided in van der Hoek & Petre (2014) contains a number of papers that consider issues related to sketching.

The paper by Mangano et al. (2015) provides an excellent discussion of the different roles of sketching in its introduction, and also provides a valuable bibliography in its references. Given the lack of an authoritative systematic review on this topic at the time of writing, this paper probably offers one of the best overviews available.

It is worth noting that, while there have been some useful laboratory-based studies in recent years, generating much of the source material referenced above, there is a distinct shortage of observational studies, and longitudinal case studies in particular. That isn't to imply that the laboratory studies have less value, and indeed, researchers have taken care to establish that what occurs is typical of designers' experiences. However, it may well be that longer-term observational studies could uncover new and richer seams of knowledge about this fascinating topic.

Key take-home points about sketching

The informal nature of sketches means that there are no rules! However, here are some potentially useful suggestions about performing sketching of design models.

Syntax and semantics. These should be treated lightly—a sketch is there to help formulate ideas, and (possibly) to help communicate those ideas to others. So words can be used to replace syntax and semantics when interpreting a sketch.

Notes. These should be used freely, and use of more 'structured' forms such as lists and tables too.

User stories. These can help with mental execution and can themselves be sketched (these latter sketches may then be the basis for the 'story boards' we discuss in a later chapter).

Recording. Keep sketches (ideally, date them too) so that they can form part of the design history, particularly where they incorporate design decisions. The rationale for the latter is easily lost, and preserving that design knowledge can be important for future evolution of an application.

Chapter 9

Modelling Software Processes

There are many needs that can well be met by applications that are organised as processes. So, having considered how we can describe different properties and attributes of software, both formally and informally, we now bring together ideas about architectural style and modelling to consider how we can model the characteristics of an application that is to be implemented as a process.

This chapter provides a fairly generic summary of how to model the different aspects of this relatively simple form of software element, since processes can be organised and implemented in different ways, and can be realised in a range of architectural styles. Indeed, an important thing to remember is that a distinctive characteristic of software is that a 'one size fits all' approach is definitely inappropriate, and that the choice of modelling forms to use needs to be adapted to the particular form of ISP being addressed.

9.1 Characteristics of software processes

In order to model software processes, it helps to have a fairly clear idea of just what is meant by a 'process'. From the perspective of an operating system, a process can be considered as being an executable block of code and associated data, that is created from some user-owned software application and is executed as a single thread of control. This interpretation can be seen as being closely related to the way that the underlying computer operates, and so historically, was a form widely used in early computing applications, and is still widely used today.

At its most simple, and most closely related to the structures of the underlying processor, this can be implemented by using a *call-and-return* architectural style. This interpretation also fits quite well with such architectural forms as *communicating-processes*, where an application might be formed by using multiple processes, perhaps operating in a *pipe-and-filter* manner, with each performing its task and then passing the outcomes on to the next one. However, since the individual processes in a communicating process structure may well be implemented by using call-and-return, we will focus on modelling this type of process in the rest of this chapter.

The key point about a process is that it has a single thread of control. Execution of a process begins with the first instruction, and follows a single sequence or path through the instructions. That sequence may be different each time it executes of course, depending on data and other factors, but for the 'main' program element, and for any sub-programs, execution will always begin at the first statement.

Figure 9.1 shows a very simple illustration of this for a program made up from a main body and two sub-programs (A and B). On this occasion, sub-program A is invoked twice, the second time exiting from an intermediate point, with the thread of control, shown as a dashed line, forming a single path.

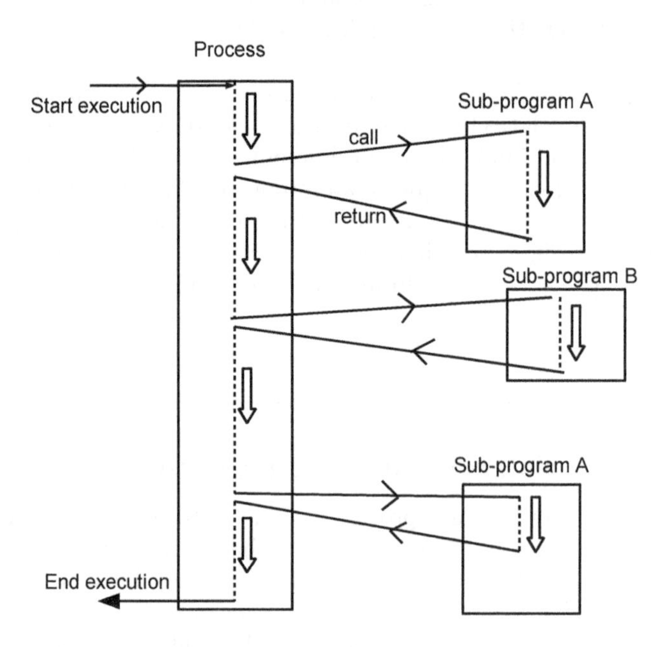

FIGURE 9.1: Simple illustration of a single thread of control

So, how can we model the way that this is organised in the abstract? The *functional* viewpoint can be used to describe the task that the process is performing, and the *constructional* viewpoint can be used to indicate how the tasks and sub-tasks involved are organised, as well as any issues related to data access. Additionally, the *behavioural* viewpoint may also aid with modelling how the process interacts with the external world, and the *data modelling* viewpoint can be used where there are important relationships in the data that may have an influence upon the structure of the process.

The issue of data organisation may be rather implementation-specific. For many programming languages permanent data storage can only be provided where variables are declared in the 'main' body of the program, with variables declared in the sub-programs only being created when that sub-program is executing. This means that knowledge about the structure, format and values of any data involved will be global in nature and shared among the elements of a process.

A disadvantage of this sharing of knowledge about data is that it forms a *technical debt* that can impede later evolution of an application. This issue was demonstrated by David Parnas (1972), with his ideas later being extended in (Parnas 1979). His crucial insight that systems constructed around *information hiding*, whereby the detailed form of data elements was only known to a few key parts of a system, made it easier to change software was a very important one, and one that underpinned the emergence of the object-oriented paradigm. (Of course, there is a trade-off, in that using this approach can be expected to result in a more complex set of structures for the organisation of the process, in order to maintain this concealment of detail.)

While information-hiding is associated with the object-oriented paradigm and the concept of *encapsulation*, it is worth noting that in (Parnas 1972) the example solution was presented as a top-down design for a process. So constructing single-threaded processes around information hiding is certainly possible, but few procedural programming languages really provide explicit support for its use.

9.2 Modelling function: the data-flow diagram (DFD)

The design of processes commonly begins with the *functional* viewpoint, since this fits well with the idea of a single thread of control that involves specifying *what* the process is to do. Modelling the *functional* aspects of a process has a domain-oriented element and hence has clear links to *requirements specification* activities. One way of describing function is to do so in terms of the way that different actions performed by the process interact with the various forms of information (data) that form a necessary part of the task of an application. Hence, what is usually termed the *data-flow* perspective, mixing actions and data, is one that has provided a highly effective way to describe the functional viewpoint for many different domains.

The data-flow approach to modelling function probably long pre-dates the use of digital technology, and it is thought such forms may well have been used in the 1920s to model the way that teams of workers in businesses were organised when performing their tasks (Page-Jones 1988). This may or may not have been the case (the evidence seems to be largely folklore), but the point remains that data-flow forms can be used to model many data-driven activities performed by people (such as processing insurance claims, or assembling flat-pack furniture) just as effectively as they can be used for modelling the activities performed by software and their interactions with data.

The name might be thought to imply that a DFD is primarily concerned with the *data-modelling* viewpoint, but the real issue here is describing the operation of a system from the perspective of the *transfer* of information, rather than being concerned about its form. DFDs are not really concerned with modelling the form of the data, but the fact that they are concerned about how it is *used* does suggest that they do incorporate some element of the data-modelling viewpoint.

When we look at *plan-driven* approaches to creating design models in Chapter 13, we will see that DFDs formed a popular starting point for many early approaches, ultimately mapping on to call-and-return implementations (Wieringa 1998). However, that doesn't mean that they can't be used with other architectural forms or development strategies (for example, they fit quite well with service-oriented architectures). In particular, where an application is strongly data-centric, they have the benefits of being:

- easy to sketch and comprehend (so, handy for describing 'user stories');
- useful for clarifying *what* an application should do, and identifying any *dependencies* upon other processes or information that this may involve.

DFDs can also be drawn with a range of degrees of formality. The 'bubble' form we use here, as popularised by Tom De Marco (1978) is relatively informal and easy to sketch. This is partly because it also makes good use of different shapes to differentiate between the elements making up a diagram (Moody 2009). In contrast the syntax employed for DFDs in the SSADM (structured systems development and analysis method) (Longworth 1992), is rather more formal and 'documentation-oriented' (see Chapter 14).

While DFDs are often thought of as being used to describe complete applications, there is really no reason why they should not be used more selectively to help clarify and understand elements of an application or its requirements. Regardless of the scope of a DFD, the top level of this is normally referred to as the *context diagram*, consisting of a single 'bubble' linked to a set of external sources and sinks of information. This is illustrated in Figure 9.2, which shows a context diagram for the operation of withdrawing cash from a 'hole-in-the-wall' dispenser.

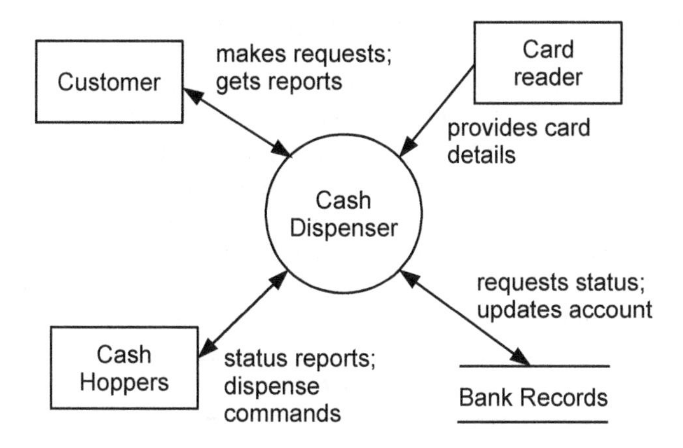

FIGURE 9.2: The context diagram for a cash withdrawal

This illustrates the four main symbols used in a DFD: the circle or 'bubble' denoting an operation; an arc indicating data flow (usually labelled with the form of data involved); parallel bars to show the use of some form of data store; and a box used to indicate the involvement of an external source or sink of information.

Turning now to the CCC, the box below shows the first steps in modelling a client request to book a car.

Reserving a car—the context diagram

The figure below is something of a mix between a formal context diagram and a sketch. The focus in this first cut at a context diagram is on determining the information that is needed, with less emphasis upon its actual source. Having identified what information is needed, it is then possible to begin thinking about how it will be acquired, as well as what will need to be output.

CUSTOMER LOCATION
RESERVE A CAR
MEMBERSHIP RECORDS
CCC ACCOUNTS RECORDS
SET OF AVAILABLE CARS

In expanding the context diagram to produce a more detailed and comprehensive model, we need to ensure that the next level of diagram will use the same inputs and outputs (of course, in the spirit of ISPs, it might be that in thinking about this, the designer realises that the context diagram doesn't cover all issues). Figure 9.3 provides an expanded DFD for the process of booking a car. Note that control issues are not described, so where, as in Bubble 5 `reserve car`, there might be two possible outcomes, depending on whether or not the chosen car has been booked out to someone else while the user is making their choice, we just draw all possible outcomes and use the text in the bubble to indicate that a choice occurs.

> *Experts focus on the essence. PvdH #37*

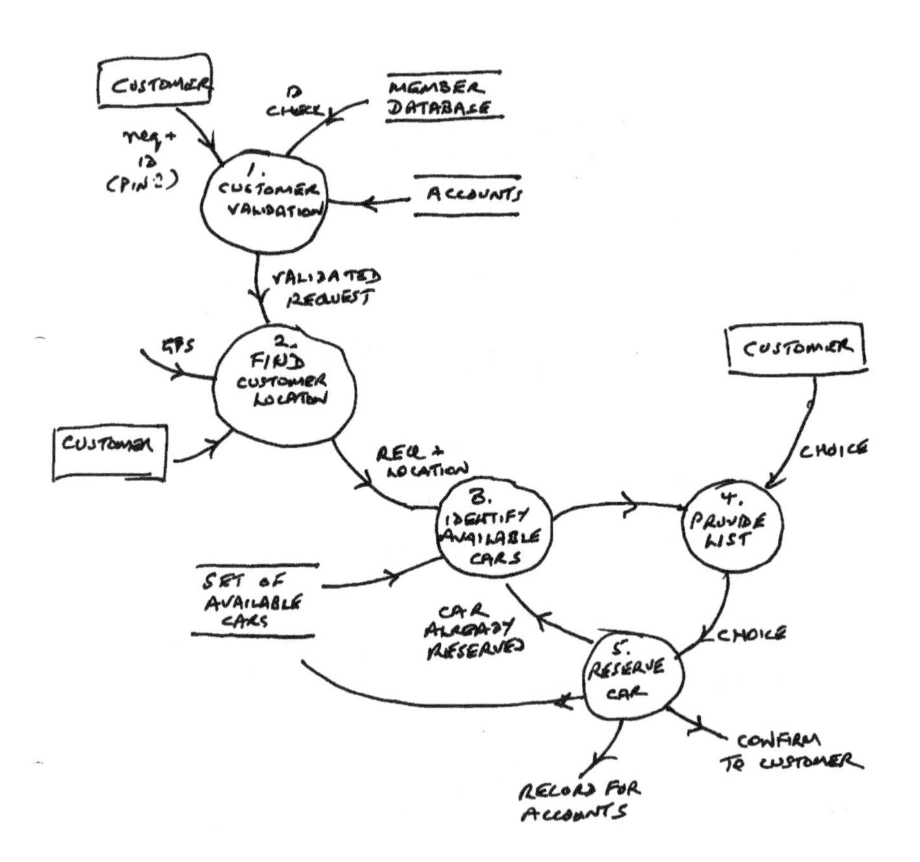

FIGURE 9.3: An expanded DFD for booking a car

(There is also the possibility that no car is available. This option can be considered as being included under `confirm to customer`, but it is poor practice to make such actions cover compound issues. So there should probably be another exit bubble to cope explicitly with the possibility that either no cars are available in the vicinity, or that there are no cars available by the time that the customer has made a choice—the reader might like to extend the diagram to handle this.)

A strength of DFDs is that they are intrinsically *hierarchical*. We can take one bubble (for example, `3. Identify available cars`) and expand this as a further, more detailed DFD. The bubbles in this new diagram will be numbered as 3.1, 3.2 etc. Being able to do this helps with managing the cognitive load of thinking about function, since the designer only needs to consider a limited set of actions at any point. At the same time, the resulting 'tree' of diagrams aids navigation around the design model at different levels of detail.

> *Experts solve simpler problems first. PvdH #2*

When developing a DFD, it is useful to start by modelling the 'physical' world (in our case, cars and customers) and then later seek to describe this more abstractly in terms of system functions as we move on to consider mapping the model on to software elements. (The corresponding DFDs are referred to as 'physical DFDs' and 'logical DFDs' respectively.) And as a final comment upon creating DFDs for the present, it is worth observing that while layout can be a bit of a challenge (as with any such diagram), the same philosophy should apply as when sketching in general. Don't worry if the diagram has things like crossing flow lines—if the diagram does become too messy and this begins to obscure the ideas it incorporates, it may be worth redrawing it, but otherwise, leave this task until later.

Figure 9.3 emphasises the issue of there being a single thread of control. For any booking action there will be a single path through the DFD. The DFD can also be used with *scenarios* of use (Ratcliffe & Budgen 2001, Ratcliffe & Budgen 2005) that define the conditions for specific execution paths, and which can also be combined with *user stories* both for developing and validating the DFD.

DFDs are easy to sketch, and as in the examples above, they do not necessarily need to be large and complex. They also provide a tried and tested way of thinking about the *functions* of a process. However, they implicitly assume the widespread availability of knowledge about the data, and so are less suited to modelling the encapsulation of data needed with object-oriented architectures.

9.3 Modelling behaviour: the state transition diagram (STD) and the state transition table (STT)

The idea of *state* is quite a familiar one. As an everyday example, we think of ourselves as being in a `sleeping state` when we are asleep, and in an `awake` state when we are not sleeping. And there are events causing transitions between these that are familiar to us, such as that caused by an alarm clock, or listening to a dull speaker in a warm room! We might also identify some other related states such as `day-dreaming` or `dozing`, which can be considered as being sub-states of a major state (in this case *awake*). This is shown as a simple model in Figure 9.4.

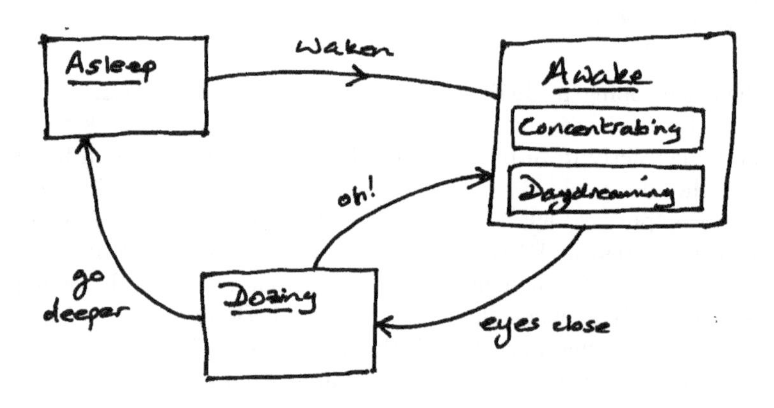

FIGURE 9.4: An everyday state model

Within computing, some classes of problem (and solution) can usefully be described and modelled by treating them as a 'finite-state machine'. Such a model can be considered as one that describes a running application as being in one of a finite set of possible states, with external (and internal) events providing the triggers that can lead to transitions occurring between those states.

A process (as well as a data element or object) in a computer fits this quite well. In a formal sense, the 'state' of a process at any point can be fully described in terms of the contents of any variables that it uses, and the 'program counter' that determines which instruction is to be executed next, although we usually prefer to use rather more abstract descriptions of state. And using a finite-state form of description to think about the properties of a process enables us to model the 'rules' which govern its behaviour. Indeed, an important aspect of such models is that they not only describe the transitions that are allowed to occur, but also those that should not be able to happen.

And of course, as a general constraint, an entity can only be in one state at any time.

We have already seen an example of state modelling when describing the different states that one of the cars owned by CCC can be in. A car can be `available`, `reserved`, or `unavailable`. This is quite a simple model, and doesn't really allow us to describe some of the situations that might arise, such as when a car is unavailable because it is damaged, or needs servicing. While that can be considered as being `unavailable` as far as modelling customer activities is concerned, it isn't really sufficient to meet all of the needs of the CCC. (And, as we saw when considering the DFD for booking a car, what should happen when two customers are offered the same car, when it is near to both of them?)

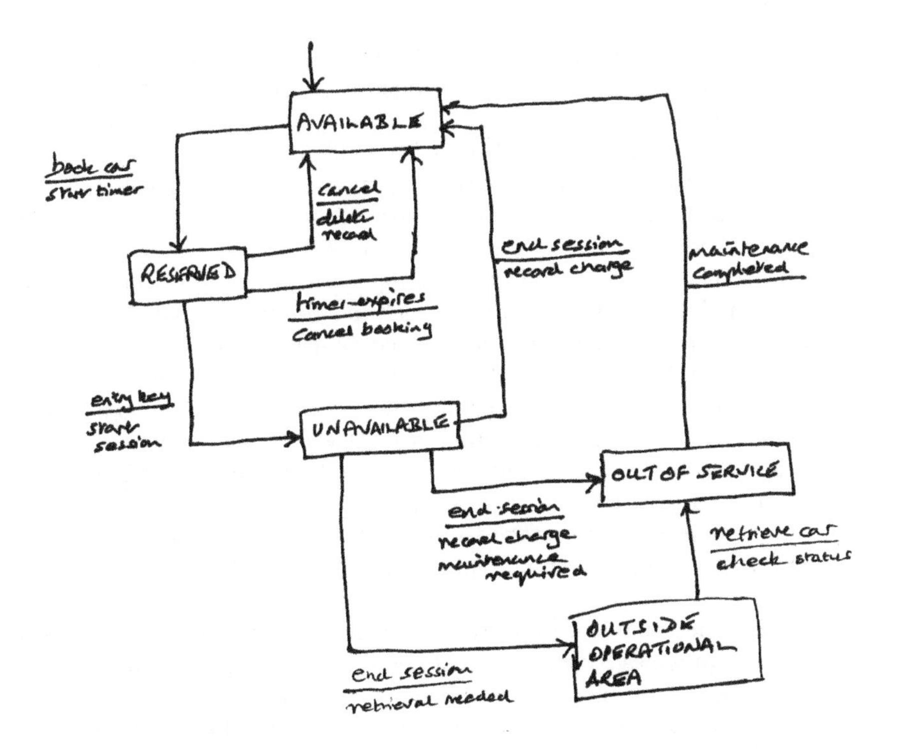

FIGURE 9.5: An extended state model for a car as an STD

Figure 9.5 shows an example of a *state transition diagram* that provides a slightly more comprehensive model that describes the different states of a car. It uses the notation developed by Ward & Mellor (1985) in which there are four principal components.

- A *state* represents an externally observable mode of behaviour for some entity, and is represented by a box which is labelled to indicate that behaviour.
- A *transition* is described by an arrow, identifying a 'legal' change of state that can occur within the entity.
- The *transition condition* that identifies the condition(s) that can lead to a transition are written next to the arrow, above the line.
- The *transition action* is written by the arrow, but below the line, and describes any actions that may occur as a result of the transition. There may be several of these, and they might occur in a sequence or simultaneously.

This model has two states that are additional to the original set, reflecting the possibilities that at the end of a session a car may require maintenance (perhaps because a routine service is due, or simply because the fuel level is too low); and that it might have been left outside of the area covered by the CCC, and require retrieval. Both of the new states can be considered as being forms of `unavailable`, a point we will return to later.

We have modelled the car as an STD in this instance, but could equally well model something 'active' such as an application or process. The value of using state modelling is that it creates a *behavioural* model that can be used to complement the *functional* model provided by using a form such as a DFD. For example, a state model can help with clarifying the 'rules' that determine which choices can be made when the process executes. There are no hard and fast rules about how and when to use an STD when modelling processes. It may be useful to begin by creating a very abstract 'system-wide' STD to help thinking about the functional tasks that the process needs to perform. Or it may be helpful at a later stage to use an STD to focus ideas about how some aspect of the application needs to behave (such as the example above of modelling the car).

Modelling the behaviour of a process (or more likely, specific elements of it) in this way provides both a means of augmenting the design model as well providing a means of performing consistency checking that all options have been considered. Essentially it is supplementary, since it doesn't usually lend itself to helping with developing the constructional model in the way that occurs with a DFD (we discuss this more fully in Chapter 13). In particular, STDs are not hierarchical in form and so for practical reasons, they are likely to be most effective when used to model elements of an application's task.

Experts test across representations. PvdH #54

Figure 9.6 provides a further example showing a rather more complex (but still incomplete) model, in this case for the behaviour of an aircraft forming

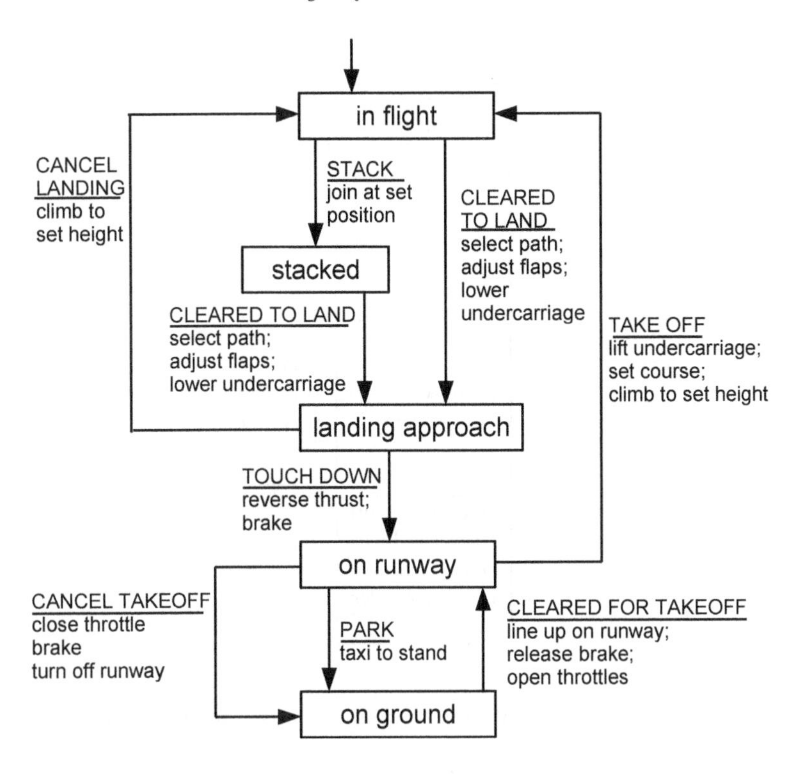

FIGURE 9.6: An example of an STD modelling aircraft behaviour

part of an air traffic control system. The short arrow at the top indicates the *initial state* (when the aircraft first enters the airspace and is detected by the primary radar). There are two sources of complexity here: one is the number of actions that an aircraft might take (including flying through the airspace and onwards, being stacked etc.) and the other is the number of operations (actions) performed in each transition (this is simplified here of course).

While STDs provide a useful visual description of how entities in a model change their state, and the operations that are involved in those changes, as well as being easily sketched, the lack of a hierarchical decomposition means that they can rapidly become inconveniently complex in form. An alternative, but less visual, way of presenting this information is to use a table, known as a *state transition table* or STT. A common convention for these is to plot the set of states down the left-hand column, and then the set of events as the column headers for the remaining columns. Entries in the table can describe both the actions to be performed and also the final state that results when a particular event occurs for a given initial state.

As an example, the model in Figure 9.5 is shown in tabular form in Table 9.1. While a *state transition table* provides essentially the same model as an STD (or can be used to do so), it can more easily be used where it is necessary

TABLE 9.1: An STT for the model of a car

	Make booking	End of session	Entry key used	20 minute time-out
Available	Record the booking; start 20 min timer; change to *reserved* status			
Reserved	Refuse booking	Cancel booking and return to *available* status	Record start time of session and location; change to *unavailable* status	Notify user of cancelled booking; change to *available*
Unavailable	Refuse booking	Record end time and location; change to *available* mode		

to handle issues of scale, or of many possible transitions between a relatively small number of states.

Additionally, the issues of verification and validation can be more systematically addressed through analysis of an STT ('are we building the system right' and 'are we building the right system') (Boehm 1981). An STT makes explicit those situations where we do not expect there to be a response to a specific event, since these correspond to an empty cell, and even as simple an act as checking and justifying all empty entries can be a useful form of analysis. In particular, when there is a need to discuss these with the 'customer', tabulation may be easier to use than diagrams—software engineers draw diagrams all the time, but others might be less comfortable with their use.

9.4 Modelling data: the entity-relationship diagram (ERD)

The *data-modelling* viewpoint also plays a rather subsidiary role when modelling processes, except of course, when a process is an element in an application that has an overall data-centric style. A form that is widely for modelling the relationships that exist between static data elements in a specification model or a design model is the *entity-relationship diagram* (ERD). While the entity-relationship concept has provided an essential foundation for the development of the models that underpin many database systems, it can help with modelling detailed data models for less data-centric applications too (Page-Jones 1988, Stevens 1991). As we will see in the next chapter, the form of the ERD has also provided useful ideas for modelling object relationships.

As with all of the notations covered in this chapter (and elsewhere) there are various, largely syntactic, variations in the way that ERDs are presented visually. However, many forms seem to have been derived from the pioneering notation devised by Peter Chen (1976). Here we concentrate on the essential elements, avoiding the more detailed nuances of the form.

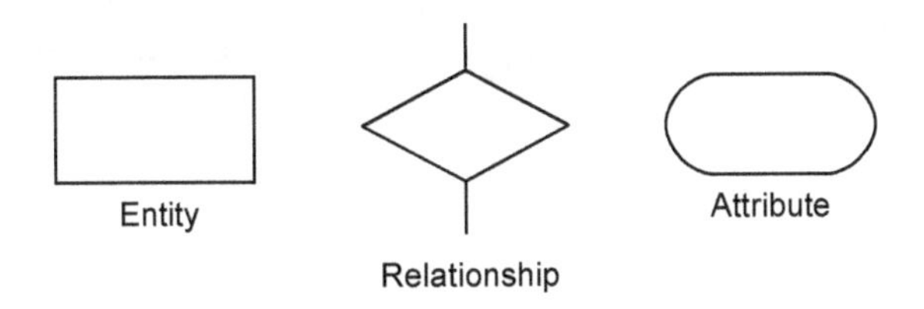

FIGURE 9.7: The basic entity-relationship notation

Figure 9.7 shows the three principal symbols that are used in ERDs, together with their meanings (the symbols for entities and relationships are fairly universal). These entities are defined as follows:

- *entities* are real-world 'objects' that have common properties;
- a *relationship* is a class of elementary facts that relates two or more entities;
- *attributes* are classes of values that represent atomic properties of either entities or relationships (the attributes of entities are apt to be more readily recognised than those of relationships, as can be seen from the examples).

We might usefully note that the ERD, like the DFD, makes good use of visual differences to clearly distinguish between the symbols.

The nature of an entity will, of course, vary with the level of abstraction involved within a design model. Entities may also be connected by more than one type of relationship. (For example, the entities `student` and `teacher` might be connected by both of the relationships `attends-class-of` and `examines`.) Also, attributes may be composite, with higher-level attributes being decomposed into lower-level attributes. (As an example of this, the abstract attribute `course-module` might be decomposed into `module-number`, `module-title` and `learning-level`.)

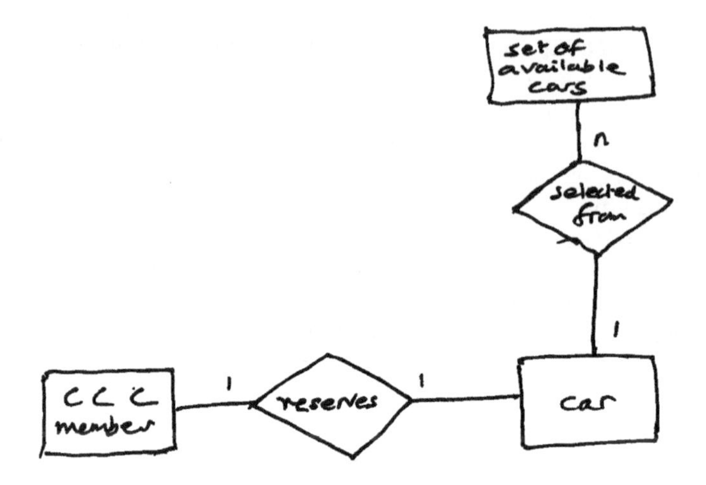

FIGURE 9.8: Some entity-relationship links in the CCC

Relationships are also classified by their 'n-ary' properties. Binary relationships link two entities. An example of a binary relation is the relationship 'reserves' that will exist between the entities `CCC member` and `car` (it isn't physically possible to drive more than one car at any moment!). Figure 9.8 shows some simple relationships that exist in the CCC. Relationships may also be 'one to many' (1 to n) and 'many to many' (m to n). Examples of these relationships are:

- `car` (of order 1) selected from the `set of cars` (entity of order n) provided in a city
- `authors` (n) having written `books` (m)

(In the latter case, an author may have written many books, and a book may have multiple authors.) The effect of the n-ary property is to set bounds upon the *cardinality* of the set of values that are permitted by the relationship.

The development of an ERD typically involves performing an analysis of specification or design documents, and classifying the terms in these as entities, relationships or attributes. The resulting list then provides the basis for developing an ERD.

Figure 9.9 provides an ERD model for the entities that might be involved in an air traffic control system (following on from the example in the previous section). This example can be considered as being quite design-related, and provides supplementary information about the factors that need to be considered in the eventual design model. The relationship between `aircraft` and `runway` also provides a simple illustration of a point that was made earlier, concerning the possible existence of multiple relationships between two entities.

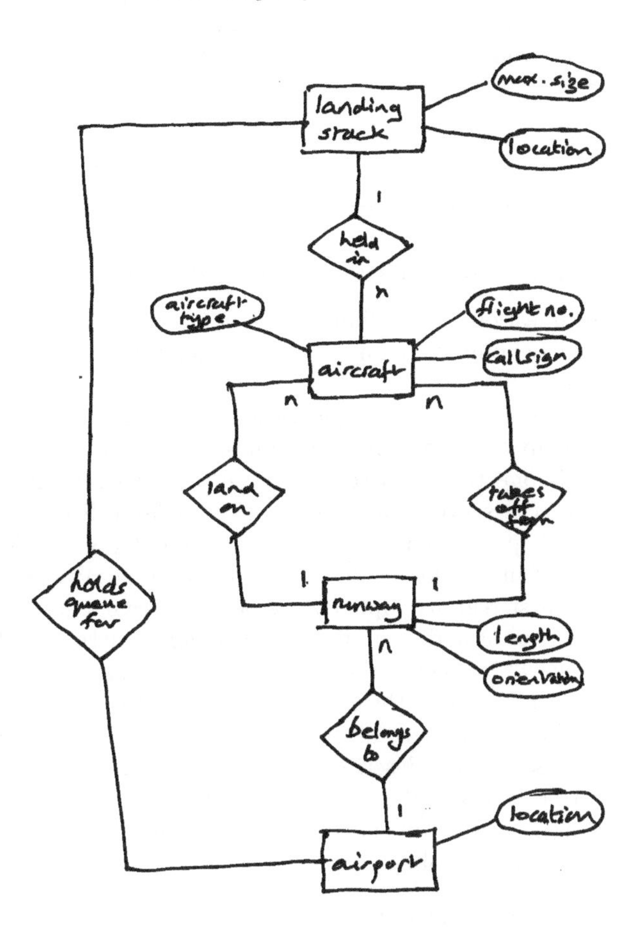

FIGURE 9.9: An ERD relating some elements of an air traffic control system

Figure 9.9 does also highlight the benefit of using visually distinctive shapes. Even though this is drawn by hand to emphasise the sketching issue discussed earlier, the symbol shapes are easily recognised.

Although ERDs are widely used for developing the schema used in relational database modelling, as we can see from this example, we can also make use of ERDs to provide supplementary elements of a process model. This is particularly relevant where the process is concerned with managing resources as in the case of air traffic control—and to a lesser degree, in the CCC.

As with the case of STDs, it may be useful to develop ERDs at different stages of designing an application. A fairly abstract model, such as that in Figure 9.9 might usefully be formulated early in the design process, and might also help clarify requirements (understanding of the IST). At other times, an ERD may be used to clarify the rules that determine how some element of the problem is to be used and changed.

Experts use notations as lenses, rather than straitjackets. PvdH #21

9.5 Modelling construction: the structure chart

We address the question of describing the structure of a process last, since the *constructional* viewpoint is commonly used to describe the outcome from modelling processes.

The *structure chart* provides a simple visual description of the hierarchy of modules making up a process. In a call-and-return architecture, the hierarchy concerned is one that is based on *invocation*, whereby higher level sub-programs invoke the services of others at a lower level. Structure charts originated in research performed at IBM to understand the problems that had been encountered in developing the OS/360 operating system, which in many ways was the first real attempt to develop large-scale software. One of these problems was that of understanding the complex structure of the code, and the structure chart was one of the forms suggested as a means of aiding understanding by visualising how the code was organised (Stevens et al. 1974).

The structure chart uses a tree-like notation to describe the hierarchy of sub-programs, and is sometimes described as a *call graph*. In terms of *coupling*, it describes a dependency based upon *control*, although some information about data coupling is usually included (there are a number of ways to do this). It uses a small set of symbols, namely:

- the *box*, which denotes a sub-program;
- the *arc*, which denotes invocation;
- some form of notation for the parameters, this might use small 'couples', which are arrows drawn at the side of the arc, or a table listing the parameters for each sub-program.

One advantage of using a table for the parameters is that this can also be used to list the global variables (usually forming part of the main body) that each sub-program uses. (When discussing the idea of *cognitive dimensions* in Chapter 7 the presence of global variables was identified as a potential problem of *hidden dependencies*, since global variables were not easily shown on a diagram). Making this data coupling explicit is useful in itself, and the use of a table also avoids having the diagram cluttered with detail. Figure 9.10 shows a simple example that uses this form and also illustrates some of the layout conventions commonly used.

Layout conventions help visualise structure, although they can get tricky with larger trees. The conventions used here include the following.

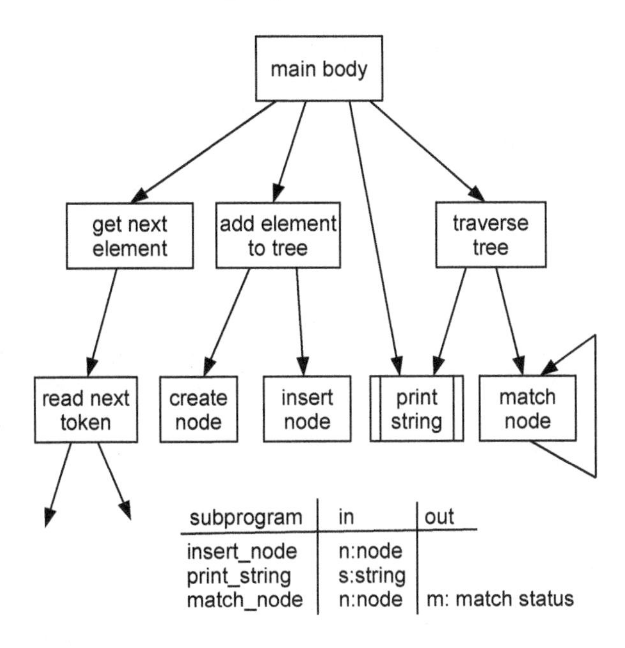

FIGURE 9.10: Simple illustration of a structure chart

- Sub-programs are grouped on levels, and each one is drawn below the level of the lowest calling unit. In the example, `print_string()` is drawn at the lowest level, because it is called from both `main()` and also `traverse_tree()`, which is on a level below `main()`.
- Double bars at the sides of a box (using `print_string()` as the example again) indicates where the designer is intending to use a standard 'library' component.
- The use of recursion can be indicated by a closed invocation arc, as in the example of `match_node`.

While there may be no explicit convention about left to right ordering of sub-programs, there may well be an implicit one. Structure charts are often drawn with input activities on the left, and output activities on the right, which probably does help with understanding of a diagram.

Because the structure chart describes an invocation hierarchy, it is implicitly hierarchical in form, and so in principle at least, any box in the diagram could be expanded using the same form. However, for moderate-sized applications at least, this is normally only likely to occur for the lowest level, as shown here for `read_next_token()`.

That said, given that some sub-programs are present purely for 'house-keeping' roles such as initialisation, rather than playing a role in the main function of the process, it may be useful to simply abstract the description of these as a single box (perhaps labelled as 'initialisation activities'), particularly as they will usually only be invoked when the process starts. Doing so makes it possible to concentrate on those sub-programs that are involved with the main purpose of the application.

Experts draw what they need and no more. PvdH #26

The structure chart does provide a relatively low level of abstraction from the final implementation of a solution. Because of this, tools do exist for reverse engineering such diagrams from code, so providing a useful tool for the maintainer. However, such tools will not readily 'group' sub-programs that are involved in performing a specific task, so layout may present a problem when analysing large applications with many sub-programs.

9.6 Empirical knowledge about modelling processes

There are few empirical studies that explicitly address topics related to the modelling of single-thread processes, and the notations employed for this. Some studies do compare forms like the ERD with object-oriented notations (usually favouring the ERD notation) and these are discussed in the next chapter. Indeed, ERDs, and the different notations used for them, have formed the basis for comparative studies such as that described in (Purchase, Welland, McGill & Colpoys 2004).

One paper of relevance here is that by Moody (2009) which examines visual notations used in software engineering. One observation from this paper is that "*most SE notations use a perceptually limited repertoire of shapes*". The De Marco form of DFD we examined in Section 9.2 is cited as an example of design excellence from this perspective. There is also some discussion of the 'Principle of Cognitive Integration', which relates to bringing together information from different diagrams (both diagrams of the same form and also diagrams of different forms). Interestingly, DFDs again demonstrate some good properties here, unlike the notations associated with the UML that we discuss in the next chapter.

Indeed, Moody's observation that "*SE visual notations are currently designed without explicit design rationale*" and that some older notations are "*better designed than more recent ones*" is a rather telling comment on the set of diagrammatical notations that together form a major design tool for software engineering.

Key take-home points about modelling processes

Modelling the attributes of processes uses a range of different forms to address the main viewpoints for each type of design element.

Design models. Modelling of single-thread processes largely uses notations that describe the *functional* and *constructional* viewpoints, although these can usefully be augmented by *behavioural* and *data modelling* notations. However, the latter are used in a supplementary role, being essentially unsuited to developing complete models.

Notational forms. While the DFD (in the De Marco format) makes good use of visual discrimination between different types of element, other notations tend to mainly use boxes and are dependent upon supplementary textual information.

Model integration. There is little scope to integrate design model information across the notations representing different viewpoints.

Tabular notations. While diagrams offer visual expressiveness, they have limited ability to handle large-scale cognitive issues. For some notations this may be aided by having a format that makes it possible to utilise hierarchy of diagram elements, but where this is lacking (as in the example of STDs) it may be useful to employ a tabular form to describe relationships in the model. Tabular forms also provide good support for checking a model for completeness and consistency.

Chapter 10

Modelling Objects and Classes

The *object paradigm* is well-established as a major architectural form and it is one that is widely employed for developing software applications across a wide range of domains and that is supported by many programming languages. While undoubtedly a very versatile form for constructing software applications, objects (and the associated classes) are quite complex structures, and much more complex than the single-threaded processes that were the topic for the previous chapter. And not only are they more complex in their structure, there are also many different forms of relationship (coupling) that can occur between objects, adding to the challenges that they present for modelling.

As a further factor that may need to be considered when modelling object-oriented structures, objects can also be realised in different ways, and this occurs with programming languages as well as run-time systems. For example, Java uses a different object model to that used by C++. In particular, Java only allows the use of single inheritance (inheriting from only one class), whereas C++ permits an object to inherit from more than one class. In this chapter we take a fairly generic approach to modelling objects, but in practice this may be constrained when it is necessary to map the design model on to a particular form of implementation.

The chapter therefore begins by examining the general characteristics of objects and classes (and the distinctions between them), in order to provide a basis for the sections that follow. An awareness of these characteristics, and how they can be employed, provides an important foundation for considering how they can be modelled. Following that, we examine the different relationships that are encompassed by the concept of *coupling* between objects, and then go on to look at some of the ways that we can describe and model these relationships between objects and classes.

10.1 Characteristics of objects and classes

In order to model software objects we need to have a set of clear ideas about what makes up the *object paradigm*. Unfortunately, clarity has not always been a quality associated with the object-oriented paradigm. Indeed, the words of Tim Rentsch (1982) could be viewed as having been quite prophetic:

> "*My guess is that object-oriented programming will be in the 1980s what structured programming was in the 1970s. Everyone will be in favour of it. Every manufacturer will promote his products as supporting it. Every manager will pay lip service to it. Every programmer will practice it (differently). And no one will know just what it is.*"

(Of course, this extended well beyond the 1980s, and we could substitute a whole list of terms or phrases in place of 'object-orientated', including agile development, software design patterns, global software development, model driven development, cloud etc. What Rentsch was really observing was the way that different forms of software development practice have in turn been seen as 'silver bullets' (Brooks 1987).) Unfortunately, silver bullets don't magically turn ISPs into WSPs, although it can be argued that while objects are unquestionably complex things, they do provide the means to model the complexity inherent in software-based ISPs.

A good overview of the concept of an object was provided in Booch (1994), which includes a survey of historical issues associated with objects. The analysis by Taivalsaari (1993) used a rather different framework and examined the notion of an 'object' from five different 'viewpoints': conceptual modelling; philosophical; software engineering or data abstraction; implementation; and formal. For the purposes of this chapter however, a rather shorter description of the characteristics of objects (and classes) will be sufficient.

10.1.1 The notion of an object

When seeking to pin down the notion of what distinguishes an object, a useful starting point may be to contrast the idea of an object with that of the process that was the topic for the previous chapter. Processes are usually single-threaded, especially when implemented in a call-and-return style. While they may exhibit *state*, this is rarely readily accessible outside of the process, and, for sub-programs any variables may be transient and only exist when the sub-program is executing. There are variations on all of these of course (that's why we have so many programming languages and platforms), but from a modelling aspect, the process is a relatively simple element, and one that is essentially 'action-oriented', with structures that place relatively little emphasis upon data or information.

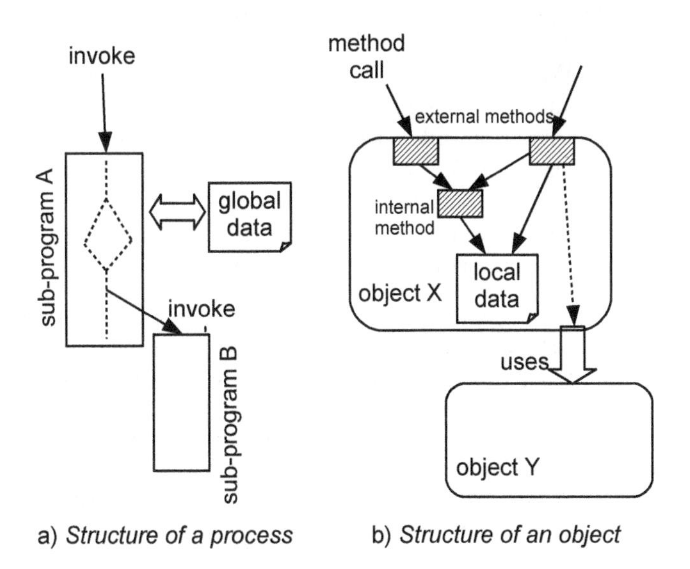

FIGURE 10.1: Processes versus objects

Figure 10.1 provides a simple illustration of the distinction between the way a process is organised and the way that an object is organised. On the left of the figure, we have a process consisting of sub-programs that make use of other sub-programs through invocation, and that can access global data structures that are held in the main body of the program. On the right of the figure, we have an object that provides multiple entry points to the resources it provides through different external *methods*, that contains permanent encapsulated data structures, employs both external and internal methods, and that can make use of other objects in a number of different ways.

As we noted in the last chapter, the emphasis placed upon 'action' in processes was challenged by David Parnas (1972) with his ideas about *information hiding*. Objects incorporate this concept via an *encapsulation* mechanism that makes it possible to create applications in which the detailed forms used to store and organise data elements are only known to a few key parts of a system.

As various ideas began to merge around the concept of objects in the 1980s, various efforts were made to maintain consistency of concepts and terminology. A useful contribution to this was made by a working group at Hewlett-Packard, for which the findings (in the form of a discussion of exactly what objects are) were reported in (Snyder 1993). A little later, the emergence of the *Unified Modeling Language* (UML) provided a set of ideas about the nature of objects that were influenced by abstract modelling. Although originally the work of the 'three amigos' (Grady Booch, Ivar Jacobson and James Rumbaugh), the UML and its evolution subsequently came under the umbrella of the *Object Management Group*[1] (OMG), and at time of writing the latest specification of the UML is that provided in version 2.5.1 (2017). It is worth noting that the OMG and those involved in the development of the UML are essentially 'computer industry' bodies, rather than the sort of grassroots organisations that have provided the motivation for *Open Source Software* (OSS) and design patterns. We will return to this issue in a later chapter.

So, we can regard an object as some form of software entity that performs computations and that has a local 'state' of some form that may be modified by the computations. In particular, an object model:

- is organised to provide *services* to other elements, rather than simply to perform actions, usually through a set of *methods* which are invoked from other objects, rather as sub-programs are invoked within processes;
- enforces strict control of scope (*encapsulation*) to ensure that data and operations within an object are not directly available for use by other elements, and can only be accessed through the external mechanisms provided by the object interface;
- makes little or no use of 'global' data, such that any data used in an application is stored within objects and normally can only be directly accessed by that object.

[1] www.omg.org

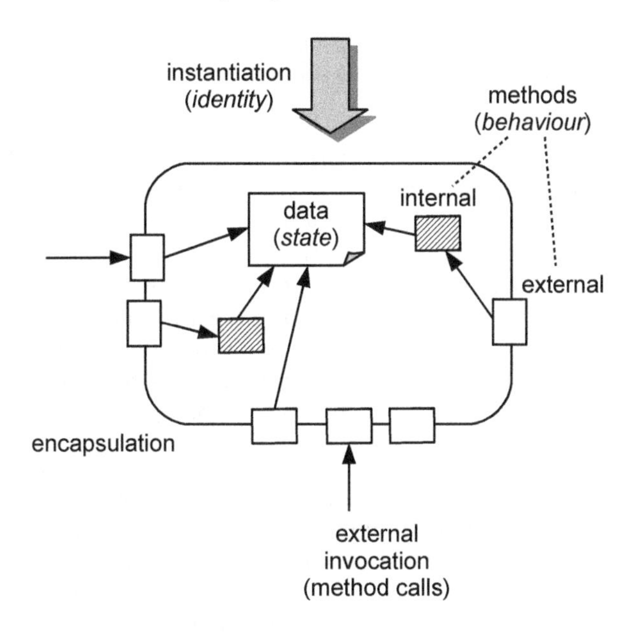

FIGURE 10.2: A simple illustration of object characteristics

An object also has a distinct *identity* that allows it to be distinguished from other objects having the same form. Hence when modelling objects, we need to be able to represent ideas about *state*, *behaviour* and *identity* in some way. Figure 10.2 provides a simple illustration of the context and form of an object.

We might also note that when modelling objects we will rarely need to model the use of global data and that the topological form of an application will usually be that of a network of objects rather than a hierarchical 'tree'. (Objects *may* have a hierarchy, in fact, they can have more than one, but not quite in the same sense that we encountered with processes, where the hierarchy was usually one of *invocation* of functional elements.)

Moving on from what an object is, to thinking about how we can model it within the design process, we can identify some key object characteristics as follows.

Abstraction This plays an important role in the design process as a whole. It is concerned with describing the external perspective of an object in terms of its 'essential features'. Abstraction provides a concise description of an object, which can then be used to help reason about its behaviour and about its relationships with other objects, without a need to possess any knowledge about its internal details.

Hence, when modelling objects, an essential property for any form of 'design object' is that it should be possible to identify its characteristics in an abstract manner. Identifying the key abstractions needed for an application and modelling their relationships is an important part of the design process.

Encapsulation The concept of *information hiding* is realised through the ability to conceal the details of an object's internal organisation and the ways in which information is represented through some form of encapsulation mechanism. Concealing the implementation details of an object makes it much easier to make changes to them without this having side-effects within the rest of the application. Encapsulation is an important issue for detailed design and implementation, and when thinking about this at a more abstract level the key question is to identify *what* should be concealed? Encapsulation and abstraction are largely complementary concepts.

Modularity This relates to the division of the overall architecture of an application into major sub-units (which we can consider as being motivated by 'separation of concerns'). In doing so, one important criterion to consider is the complexity of the interfaces between the modules, while another is the likely effects of evolution of the application over time. For processes, the unit of modularity is the sub-program, which is largely organised around function, with relatively little emphasis being placed upon any relationships with data. In an object-oriented context where there are many forms of *uses* relationships, the choice of suitable modules becomes rather more complex and multi-faceted. For example, objects do not have a single thread of control, and it may be that the design of an application needs to allow different methods to be invoked concurrently, requiring that any consequent changes to variables be suitably protected.

The choice of modules is also determined by how the application will evolve. Where possible, we should be seeking to isolate major design decisions within separate modules, so that when changes do occur, they are largely isolated to a single module. We look at this later when we discuss the role of *architectural patterns* in Chapter 15.

Hierarchy As noted above, the object model is characterised by the presence of several different forms of hierarchical structure. Within an object that performs complex computations there is the possibility of having a hierarchy of *function*, rather similar to that of processes, whereby a complex task is sub-divided into small elements. There is also the possible use of a hierarchy based upon *class* structure, which is examined more fully in the rest of this section. Finally, there is the extent to which the *interdependence* of objects forms a hierarchy, usually referred to as the *uses* relationship, and this is explored more fully in the section that follows.

Before we examine the idea of a class hierarchy, there are two other characteristics of objects that should be mentioned, since they can affect the modelling process. The first is that new objects can be *created* at any time—and obviously, objects can be deleted too. This creates a dynamically changing application structure, and is a particularly useful feature when a particular object is mainly concerned with representing one item of a set of resources and we want to add new items to the set. However, it also represents a new issue in terms of the ideas about design models that we have so far encountered, since the equivalent feature for processes is limited to creating new list elements when using linked data structures. And a second, related characteristic, is that when a client issues a request to one object in a set of objects, it needs to be able to identify the object that should be the recipient (taking us back to the issue of objects having an *identity*).

10.1.2 Objects and classes

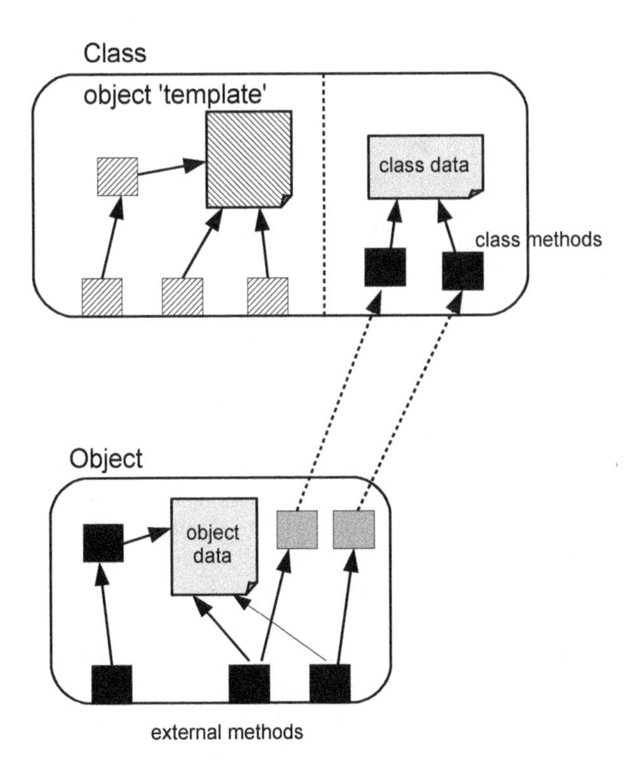

FIGURE 10.3: Instantiating an object from its class

Viewed in terms of their implementation roles, a class specification can be regarded as a form of 'template' that can be used to create objects. The class defines the state and behaviour attributes of an object, and when an object is created (instantiated) from a class, it acquires a set of variables and methods for that class. It also acquires a unique identity. However, that is not all, for depending upon how the object model is implemented, the class itself can provide some shared resources, visible to all of the objects derived from it. These may include class methods and class variables that do not form part of an object. For example, a class may keep a count of the objects created from it, using a class variable to store this, and employing class methods to increment/decrement the count. Figure 10.3 shows the class and object relationship in a schematic form. The key issue here is that the class data is essentially 'in scope' to the object in the same way as any object data.

As an example, a simple form of this could be used in the CCC to limit the number of customers allowed to request a car at any point in time. Each 'customer object' can be linked to the details of up to three 'car objects', to provide the customer with a choice. However, it makes little sense to allow the same car to be offered to many customers at the same time. One way to constrain this would be for the 'customer class' to contain a counter and a condition that constrains the number of customer objects created at any point in time to be (say) half of the number of available cars. When this limit is reached, no further customers would be allowed to request a car. (Of course there would also need to be an accompanying mechanism to 'queue' requests from customers resulting in new customer objects.)

The hierarchy that is based upon class structure also leads us to consider the concept of *inheritance*, by which the properties of an object can be derived from the properties of other objects, or more correctly, of other classes. The question of how important inheritance is for object-oriented design comes close to being a theological issue. In particular, it introduces an issue that we will return to later, as to whether (or when) the 'right' design abstraction that should be used for modelling is the *object* or the *class*. So this is a good point to discuss the class/object relationship more fully.

In a programming language it is usually possible to create new data types from existing types, particularly by compounding these in some manner, or by defining a sub-range type. We can do much the same with classes, but the associated *inheritance* mechanism used to create subclasses is much more complex than the forms needed for creating derived types, since it needs to incorporate not just static data relationships, but also behavioural qualities (methods).

The subclasses of a class will share common structures and common features of behaviour. As an example, a bank may provide customers with a choice from many different forms of account, each of which may have different rules about how interest is calculated and paid, how charges are applied, the minimum balance required, overdraft rules and so on. However, all of these forms are clearly recognisable as being subclasses of some 'parent' class of `bank_account`, and share some common information structures (characteristics), such as those used to describe:

- the current balance

- the identity of the account holder
- the date of creation for the account

as well as some common operations, such as:

- creation of a new account
- addition of interest
- making a deposit
- withdrawal of part of the balance

So, for this example, the class of `bank_account` will provide a description of these common properties and behaviour, while the detailed form of a particular subclass of account will be structured around the rules applying to that subclass. For design purposes, the designer therefore needs to concern themselves with the abstractions involved in the class (`bank_account`), and the subclasses. Figure 10.4 illustrates these points.

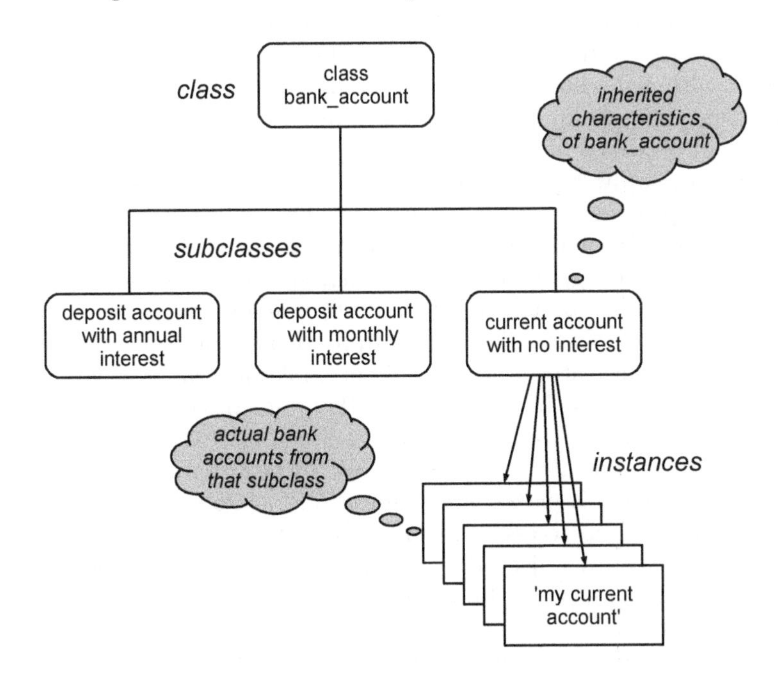

FIGURE 10.4: The inheritance hierarchy

Inheritance provides the mechanism by which subclasses acquire the general properties of their parent class(es) (also known as *superclasses*). In our example, any specific form of bank account will inherit the properties of bank accounts in general, but may have some features that only apply to that subclass (extended methods, additional methods, minimum balance etc.). So inheritance is an important constructional concept, and one that has been incorporated into many different programming languages.

The form depicted in Figure 10.4 is based upon the use of *single inheritance*, whereby a class can only inherit the properties of one superclass. (Of course, that superclass may itself inherit from a higher superclass.) However, it is also possible to inherit from more than one superclass, forming *multiple inheritance.* Obviously this does add considerably to the potential complexity involved in keeping track of the inheritance relationships. Hence many programming languages, most notably Java, only permit the use of single inheritance. From the point of modelling design ideas, multiple inheritance certainly increases the cognitive load involved, as well as adding to the technical debt by creating potential problems for future maintenance (Wood, Daly, Miller & Roper 1999).

> *Experts prefer simple solutions. PvdH #1*

A mechanism often introduced alongside inheritance is that of *polymorphism.* And like inheritance its use is apt to be associated with detailed design decisions, although conceptually at least, the decision to employ polymorphism could arise at any point. What it really relates to is the more flexible options that objects provide with regard to *binding time.*

When creating an application using a call-and-return form, the decision as to which sub-program is going to be called to perform some task is essentially fixed at the point when the code is written and compiled. The run-time process will contain the linkage information for that specific choice of sub-program, and this cannot be modified. So call-and-return architectures usually embody what we can regard as a *static* binding to particular sub-programs.

Figure 10.5 illustrates a very simple example of what can happen in an object-oriented system. In this example, there are three objects, designated X, Y and Z (which we can also assume are created from different classes). The block arrows indicate 'makes use of', so object X makes use of both other objects, while object Z makes use of object Y. Each object provides a set of methods that provide access to its encapsulated data. So, if object X needs to access data held by object Y, it might do so by using the methods `Y.b()` or `Y.c()` for that purpose. (Because objects have identity, it is a common convention to refer to a method as *object_identifier.method_identifier.*)

However, we might also note that object Z also provides a method `b()`. Since this can be referenced as `Z.b()`, no confusion need arise. However, in an object-oriented context, whenever object X makes a reference to the method `b()` with no *object_identifier*, we can expect that the context of that reference will be used to determine whether the appropriate method to use is `Y.b()` or `Z.b()`. And the decision about the appropriate choice will be made automatically and at *run-time.* This is because the bindings between objects are

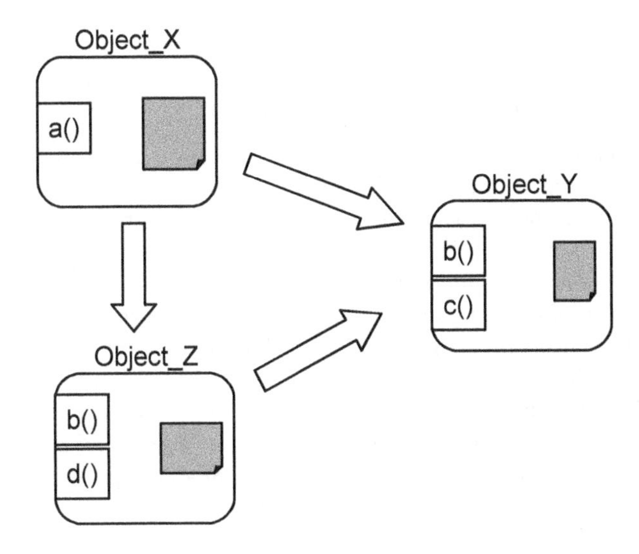

FIGURE 10.5: Dynamic binding and polymorphism

created *dynamically* when the application executes, allowing the appropriate choice to be made at that point.

It is this facility for selecting the appropriate method from a number of methods with the same identifier, but originating in different classes, that is termed *polymorphism*. (Strictly, although our example is couched in terms of objects, the methods themselves are defined in the parent classes.). To return for a moment to the earlier example of the bank account and its subclasses, we might expect that all classes will provide a `withdraw()` method, but that the detailed form of how this operates will be different, according to the type of account involved. So when this method is used in an application, the choice of which instantiation of `withdraw()` will be employed will depend upon the type of account provided as its argument. (This is also why we tend to associate polymorphism with inheritance, since inheritance provides a mechanism for exploiting this in an elegant manner.)

What is important here though is not the mechanism of polymorphism itself, but the way that it highlights the use of dynamic binding, and the characteristic that methods belong to objects, rather than being a static binding. From a design perspective, it also offers a quite radically different conceptual model.

The class-object relationship does complicate modelling and design in many ways, and not just through the use of such mechanisms as inheritance and polymorphism. Initial modelling may well be more problem-related, suggesting that the emphasis at that stage is likely to be upon objects. However, later stages of modelling, where constructional aspects become more important, may well use classes too.

10.2 Relationships between objects

In the preceding section the emphasis was upon relationships that can occur between objects (and classes) that share attributes. In this section we examine the relationships that can exist between objects that are unrelated, but that need to work together in order to produce an application with the required behaviour and functionality.

A useful way of thinking about these and other relationships is to consider the different relationships that have been found to provide useful measures that can help with profiling the form of object-oriented applications. The set of six *metrics* identified by Chidamber & Kemerer (1994) (usually abbreviated to C&K) are widely used to assess the structures of these, and have themselves been derived from consideration of the object model. The metrics, together with their common acronyms are summarised in Table 10.1.

TABLE 10.1: The Chidamber & Kemerer metrics

Metric	Label	Description
Weighted Methods per Class	WMC	The sum of the 'complexity' values for all of the methods in a class.
Depth of Inheritance Tree	DIT	A count of 'tree height' for ancestor classes.
Number of Children	NOC	The number of subclasses that inherit from the given class.
Coupling between Objects	CBO	The number of couples that exist between a class and other classes.
Response for a Class	RFC	A measure of the "immediate surroundings of a class".
Lack of Cohesion in Methods	LCOM	The number of methods in a class that do not share attributes.

These metrics provide surrogate measures for some of the object-oriented concepts (which inevitably are not easily measured directly). They are probably most usefully used for comparison, both *between* the properties of different classes, and also for the way that the properties of a particular class *change* as a result of maintenance. As metrics they are often used to help identify the classes that are considered to be most likely to contain faults. We discuss a systematic review of their usefulness for identifying modules likely to contain faults in Section 10.9, and for this section just provide brief comments that are based upon its findings (Radjenović, Heričko, Torkar & Živkovič 2013).

Two of the metrics (DIT and NOC) really belong with the discussion of the previous section, so we only need to comment here that while DIT probably offers no significant predictive ability for fault-proneness, NOC is in the group that does have "some ability to discriminate" with regard to the effect of changes to a class. NOC does give some indication of the potential influence that a class has on the design.

Two of the other metrics (WMC and LCOM) are calculated on a class basis, but are still useful when comparing values for different classes. *Weighted Methods per Class* (WMC) is calculated as:

$$\sum_{i=1}^{n} c_i$$

where c_i is the complexity of method i. The choice of the complexity measure c is left to the user, and essentially the metric calculates the overall complexity of the class as a function of the complexities of the individual elements. One choice for complexity is simply to set it to a value of 1, which produces a count of the methods in the class. Other values that are used include Lines of Code (LoC) and McCabe's Cyclomatic Complexity (which is a measure of the number of possible execution paths through a method). While WMC as a metric is of less use for making comparisons with other objects and classes, research does suggest that an increase in WMC after changes are made to a class is a good indicator of a possible increase in defects for that class.

Lack of Cohesion in Methods (LCOM) is based upon the idea that the elements of a class should be related to the purpose of the class, and hence should share some attributes. In an earlier version of the metric definitions, this was a count of methods that did not share variables, so that for a cohesive class we would expect a value of 0. This was later re-defined as follows:

$$\sum(pairswithnocommonattributes) - \sum(pairswithcommonattributes)$$

which addresses the same concept more clearly. (Note though the comment about the validity of LCOM in the section on empirical evidence.)

The remaining two metrics are very much concerned with the interactions between objects (or classes).

Coupling between Objects (CBO) is the count of the number of couples that exist between a given class and other classes. Such a *coupling* is assumed to exist if one class *uses* methods or instance variables from another class, and an excessively high value of CBO is considered to be an indicator of poor modular design as well as an impediment to reuse. Coupling can take a range of forms: invocation of a method, inheritance, copying the value of a variable. The metric treats coupling as a two-way link, and so calculation of CBO needs to count both the classes that use a class and also those that it uses, as shown in the example of Figure 10.6.

In the example, the solid lines indicate that one object calls methods in another (such as the line between class A and class B) while the dashed line between class A and class C indicates a different form of coupling, such as inheritance (class C inheriting from class A).

As often, the empirical findings about the usefulness of CBO are rather mixed, but overall, it was found to have "moderate predictive ability" with

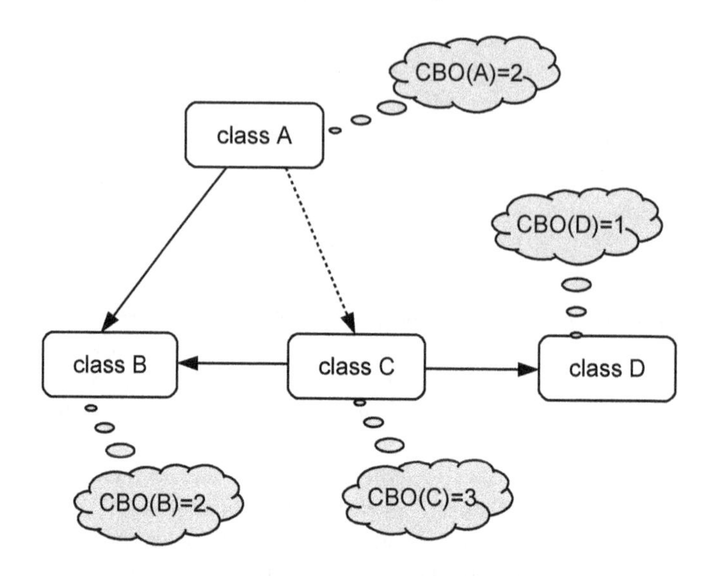

FIGURE 10.6: Simple illustration of CBO measures

regard to fault-proneness. Two useful observations from a design perspective are that:

- the presence of one class having a high value of CBO, while others have a CBO of 1, may indicate that a call-and-return design model has been mapped on to an object structure;
- the presence of many classes with high CBO may indicate that the modularity in the design model is too granular, and that the classes are too small, requiring that they have to make use of other classes in order to perform their tasks.

Response for a Class (RFC) is a measure that seeks to reflect the influence of the "immediate surroundings of a class", in other words, the methods that it uses directly. The *response set* of a class is the set of methods that are accessed by the set of methods belonging to an object of that class (which may include access to other methods within the class). This is a static measure, and so identifies the set of methods that might potentially be executed if a message is sent to one of the methods in an object of that class. It is defined as:

$$\sum(localmethods) + \sum(methodscalledbylocalmethods)$$

and Figure 10.7 shows a simple illustration of this.

As a measure, the larger the number of methods that can be invoked from an object, the greater will be its complexity in terms of the effort needed to comprehend its operation, and of the level of understanding needed to test it.

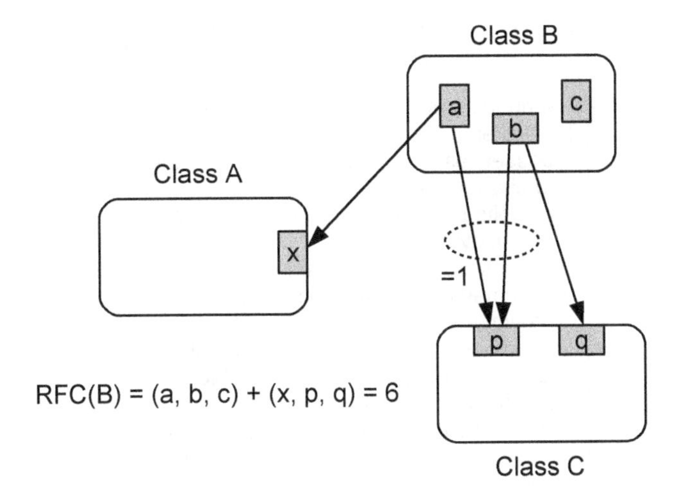

FIGURE 10.7: Simple illustration of RFC measures

There is limited research into the value of RFC, but what is available does indicate that this measure may correlate well with the likely number of defects in the class.

Of course, the things that we can measure (such as the C&K metrics) are not necessarily a guide to how we should design an application, and they are largely concerned with relatively static aspects of object relationships. However, the constructional viewpoint does play an important role in OO modelling, and so the measures do give some indication of the sort of relationships that we need to think about, and the issues that should perhaps be anticipated when producing a design model.

10.3 Conceptual issues for object modelling

Figure 10.8 summarises some of the main characteristics of objects that have been mentioned in the preceding sections (obviously, they are not the only ones). Here we briefly look at the conceptual issues that are implied by these when thinking about how we can model object-oriented designs.

To model a design solution with the goal of employing any particular form of implementation (in this case objects), the designer needs to possess not only a good conceptualisation of the relevant architectural style, but also some clear cognitive mappings that can be used to link ideas about the design model to the implementation constructs that characterise that style. For the

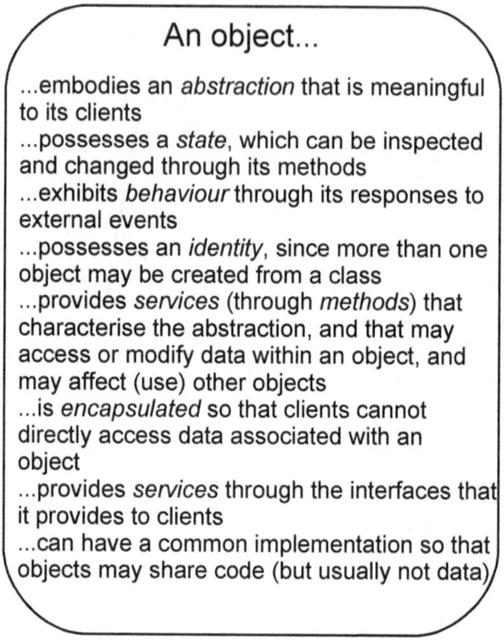

An object...

...embodies an *abstraction* that is meaningful to its clients
...possesses a *state*, which can be inspected and changed through its methods
...exhibits *behaviour* through its responses to external events
...possesses an *identity*, since more than one object may be created from a class
...provides *services* (through *methods*) that characterise the abstraction, and that may access or modify data within an object, and may affect (use) other objects
...is *encapsulated* so that clients cannot directly access data associated with an object
...provides *services* through the interfaces that it provides to clients
...can have a common implementation so that objects may share code (but usually not data)

FIGURE 10.8: Key characteristics of an object

call-and-return style, the abstract model is one that is described in terms of statically linked sub-programs, with these having well-defined functional roles. The control and data topologies are also more or less the same, so that the task of mapping a design model on to a given procedural programming language is a relatively straightforward process. Likewise, the *pipe-and-filter* style employs fairly simple architectural elements, and again, both control and data topologies are closely linked, assisting with the eventual implementation.

When employing the *object-oriented* style, the basic concepts used for formulating the abstract design model, as well as for mapping it on to some form of implementation, are somewhat more challenging. To take a simple example, we noted earlier that observational studies of designers showed that they often 'mentally executed' their design models, to ensure that these exhibited the intended behaviour. This is a relatively straightforward exercise with a design model formulated using a style such as call-and-return, but a potentially much more complex task when realising a design using objects. The differing control and data topologies, together with dynamic binding of methods and multiple threads of execution within an object, all combine to make this more challenging.

Experts know how things work. PvdH #33

One of the arguments sometimes made for adopting the object-oriented style is that it is more 'natural' than (say) call-and-return. Objects and their interactions can be recognised in the everyday world, whereas call-and-return uses a model which more closely reflects the workings of the computer itself. Hence (so the argument goes), the process of analysis should involve identifying the objects in the problem domain, and then use the resulting model to help with deciding upon a set of corresponding 'solution objects'. However, in practice this has not proved to be a particularly effective strategy, and as Détienne (2002) observes:

> "*early books on OO emphasised how easy it was to identify objects while later ones, often by the same authors, emphasise the difficulty of identifying them*"

which rather undermines the case for 'naturalness' and for any modelling based upon it. That said, Détienne also notes that comparative studies have demonstrated that:

- object-oriented design tends to be faster and easier than the procedural (structured) design approaches discussed in Part III;
- different designers will produce solutions that are more similar when using an object-oriented approach than when using other approaches;

suggesting that once the necessary levels of knowledge have been acquired by the designer, the object-oriented style may have more to offer.

An important issue here would appear to be the relatively steep learning curve involved in *learning* about object-oriented design. Studies by Fichman & Kemerer (1997), Vessey & Conger (1994) and Sheetz & Tegarden (1996) all observed that inexperienced designers struggled to acquire and deploy the concepts. In the study by Sheetz & Tegarden (1996), they specifically identified the following object-centred design issues as contributing to the problems by acting as sources of confusion for the inexperienced designer.

- Organising the distribution of application functionality from the 'problem space' across a set of objects. (In other words, identifying the objects.)
- Using the existing class hierarchy.
- Designing classes (including making use of inheritance).
- Using polymorphism, where the semantics of methods with the same name may differ.
- Evaluating solutions (see earlier comment about mental execution).
- Communication among objects.
- Designing methods.

Some of this complexity relates largely to constructional issues, but much also arises from the cognitive load that is imposed by the need to comprehend and model so many issues at each step in design. The latter aspect will become more evident when we look at design processes in Part III. For the moment, we will concentrate on how we might model object-oriented properties.

10.4 Object modelling: the issue of notations

Various forms of diagrammatical notation have been employed for modelling the different properties of objects. The best known of these is probably the UML (Rumbaugh, Jacobson & Booch 1999), mentioned earlier in the chapter. In his critique of the choices of visual notations used in modelling languages, Moody (2009) comments on the poor use of different shapes in many SE notations and observes that they "use a very limited repertoire of shapes, mostly rectangle variants" and goes on to point out that rectangles "are the least effective shapes for human visual processing" and that "curved, 3D, and iconic shapes should be preferred".

Unfortunately, most OO notations, and the UML in particular, make extensive use of rectangles. Perhaps the one thing that we can say in their favour is that they are easily drawn on a whiteboard (one of the problems with both colour and icons is that they do present more of a challenge to sketching skills).

For the purposes of the rest of this chapter, rectangles have been retained where an established notational form is being used, but different shapes have been used as far as possible. Similarly, some examples have been sketched, for the reasons explained in Chapter 8. Specific forms from the UML are used explicitly where these have no obvious equivalents (for example, activity diagrams), but wherever possible, the following sections make use of fairly general forms of notation and syntax. It is worth noting that UML 2.0 uses a large set of diagrammatical forms (13), with some overlaps. However, only a key subset of these is described here and in later chapters!

10.5 Modelling construction: the class diagram

The constructional viewpoint tends to play a rather different role for developing design ideas for objects than the one that it takes when producing design models based upon processes. In the latter case design models tend to be developed using the other viewpoints, and constructional forms are used at a relatively late stage in design (see Chapter 13). With the object model, the constructional viewpoint may well be used throughout the design process, probably because of the importance of identifying the key objects at an early stage. Early stages in design may use fairly simple descriptions of classes and objects, with more elaborate forms being used for detailed design.

Two aspects complicate description when employing the constructional viewpoint. Firstly there is the need to distinguish between a class and an object created from that class. And secondly, the designer has the added complexity of needing to model a number of quite different forms of *uses* relationship. This certainly complicates the role for any diagrammatical notation since, with relationships being 'connectors' between objects, it is likely to be more difficult to employ the sort of variety of visual shapes that can be adopted for the elements of a diagram. Equally, to understand and use a model, it is important to be aware of the nature of a particular relationship.

Classes themselves are usually modelled as rectangles. Early forms of OO design did use different shapes (for example, in Robinson (1992) a box with rounded corners was used for *Hierarchical Object-Oriented Design* (HOOD) objects. Since we spend a lot of time drawing objects when modelling, the choice of a simple rectangle is probably a practical one to employ, and it is of course easy to sketch.

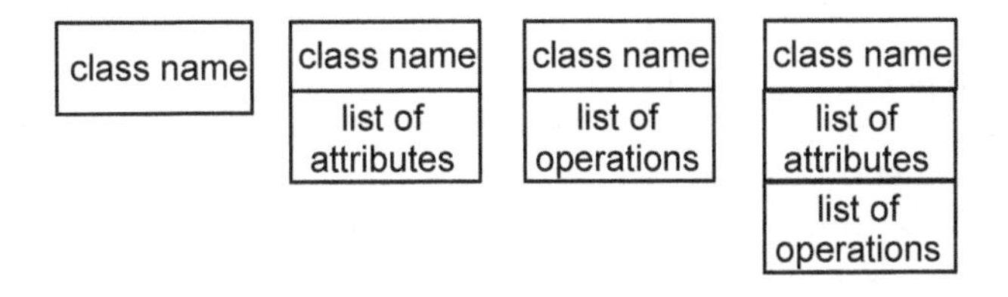

FIGURE 10.9: The UML class notation

The UML has adopted the use of a box in its *class diagram*, and classes can be represented using boxes with three compartments, two of which are optional. A class can simply be represented by a box containing the *class name*, which is probably sufficient when the class is first included in the model. The two optional compartments are the *attributes*, with the state of an object of the class at any time being determined by their values; and a list of the *operations* that the class provides through its methods. The resulting forms, using one, two or three compartments are shown in Figure 10.9. Since this is quite a flexible form, we will generally make use of it in the examples that follow.

> *Experts draw what they need and no more. PvdH #26*

We might also note that the class diagram doubles up for the *data-modelling* role, largely because objects incorporate both methods and data.

10.5.1 Distinguishing classes from objects

As explained earlier, the distinction between a class and an instantiation (object) created from that class can be quite important, particularly for those classes that may be used to create many objects. However, classes may often be used to create a single object, and so the distinction may well be one that can be ignored for much of the process of design.

However, since it can't be completely avoided, the question is then, how to make the distinction. The closeness of form between the two means that using different shapes is not really a practical way of making the distinction, and so this does tend to rely upon annotation. (Unfortunately, for the UML in particular, so do rather too many other features.)

The mechanism adopted in the UML is to give an object a name that is made up of the name of the specific object, followed by a colon, with this then being followed by the class name. The full name is then underlined, to emphasise that it is an object (this is particularly important with anonymous objects, since they simply have a name made up of the class name, preceded by a colon). So an object name (in the top compartment) may look like:

`objectName:Class`

(It is also common practice to use the *camelCase* style for the object name, which may also help to make the distinction between an object and a class. A camelCase identifier is made up of several concatenated words, with all but the first word beginning with an uppercase letter.)

Underlining is probably a pragmatic option and certainly preferable to making use of typographical features such as bold, italics or other forms that are not easily sketched on a whiteboard. Figure 10.10 provides a very simple example of a class descriptor and an associated basic object descriptor. The compartments for status information and the key methods have been included for the example of the class descriptor, while the object descriptor just uses the basic compartment containing the identifier.

The UML provides many options for supplying detail about status and operations for both classes and objects, including information about types, initial values and visibility. While possibly relevant for detailed design should it be required, these are unlikely to be useful while developing a design and

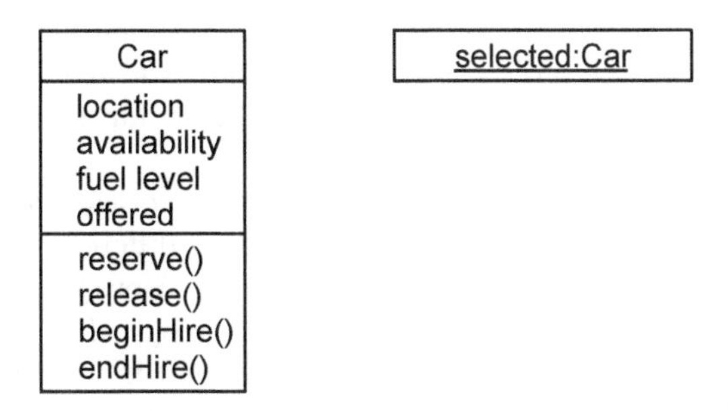

FIGURE 10.10: The UML class-object notations

so are not addressed here. More details about these features can be obtained from specialist texts on OO modelling such as (Arlow & Neustadt 2005) or (Lunn 2003).

10.5.2 Class relationships

In order to make use of the class and object notations we need to be able to model the different *relationships* that can occur between them. These are generally drawn as one or more *object diagrams* that represent the structure of an application at some point in time. At its most basic an object diagram forms a network model that describes the interactions between a set of objects, where these interactions chiefly consist of method calls. (The UML refers to the links between classes as *associations*.)

In the same way that we employ the concept of *arity* with *entity-relationship diagrams*, when modelling relationships between classes it may be useful to indicate whether a relationship between two classes is on a one-to-one, one-to-many or many-to-many basis. The emphasis here is upon *may*. If the application consists of a small number of objects, with only one object from each class, there is little point in cluttering up the model with information about the *multiplicity* of the relationship.

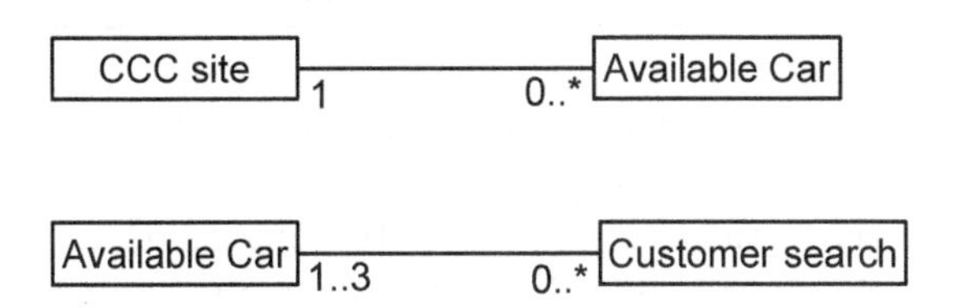

FIGURE 10.11: Use of class multiplicity annotation

Figure 10.11 illustrates the use of multiplicity for two simple examples from the CCC. The upper one indicates that a CCC site (assuming that CCC expand their business to other cities) can have between zero and many cars available. The lower one indicates that a car that is available can respond to any number of requests, but that the 'search object' for a customer may only select up to three cars.

Horrified purist

The UML has some less than inspiring notations for indicating different forms of relationship between classes. However, it also provides the quite useful concept of a *stereotype*, which effectively allows a new modelling element to be introduced that is based upon an existing one. Stereotypes are created by placing the name of the stereotype between guillemots («...») and using it to label the line indicating the relationship. Although it might horrify a UML purist, using stereotypes as a means of clarifying intended relationships, especially when sketching, may be much clearer than using the graphical symbols defined in the UML.

When describing class–instance relationships we might therefore use the stereotype «instantiate». Another useful one is «use» (this one is so obvious that it can simply be treated as a default). The role of «create» is likewise obvious. Although not advocated by the UML, it may also be useful to use stereotypes for other relationships such as «inherit» and «aggregate». Reducing the number and complexity of symbols is a useful step when dealing with objects and doing so in this way reduces the cognitive load. It also demonstrates that you don't *have* to use the formal semantics of a notation, especially when sketching ideas.

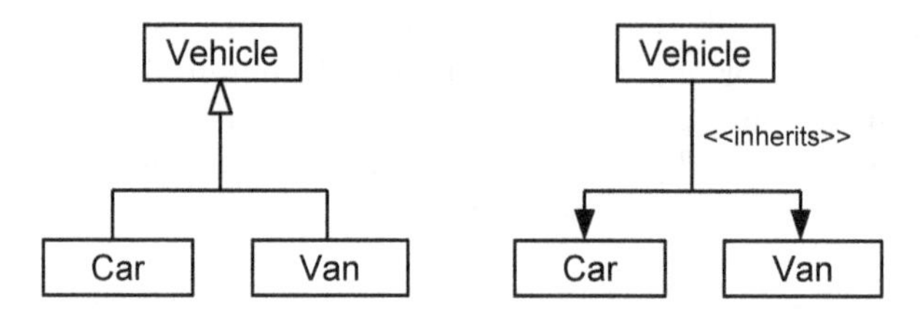

FIGURE 10.12: UML class inheritance annotation versus using stereotypes

Figure 10.12 shows the conventional UML notation for inheritance on the left, and the use of a stereotype on the right. There is definitely an argument in favour of using the stereotype for showing such a diagram to anyone unfamiliar with the UML, as well as when sketching on a whiteboard. However, one thing that can be harder to capture this way is the direction in which the 'flow' of inheritance should be read (although many might find the UML notation confusing anyway, in terms of the apparent 'flow' implied by the triangle). So careful positioning of the indexstereotype stereotype or using an additional arrowhead as shown here might sometimes be helpful.

Other forms of association, including *aggregation* and *composition* can be used when modelling classes and objects. And of course, we can add *polymorphism* and *abstract classes* to our use of *inheritance*. All of this explains why books on modelling with the UML tend to be large—as classes and objects are *much* more complex design elements than processes. So, in fairness, while the UML itself may often be cumbersome as a modelling tool, it is required to provide ways of modelling some potentially very complex structures.

10.6 Modelling behaviour: the statechart and the message sequence diagram

The idea of *state* plays a much bigger role for objects than it does for processes, not least because the use of encapsulation means that an object is very likely to contain persistent information related to its role and identity. And because an object can have many external methods, there can also be many ways of accessing and modifying that state.

Object modelling commonly makes use of two forms for modelling behaviour. The *statechart* (or 'state diagram') is used to model the way that the state of some entity (which may be an object, or a subsystem, or a complete application) is modified by interaction with the events occurring in the external world. And the *message sequence diagram*, or *message sequence graph* (MSG) can be used to model the way that events are triggered and handled over time. Actually, sequence diagrams are not entirely about behaviour, they can also be viewed as describing some aspects of *function*, highlighting the problem with any simple classification system! However, regardless of how we classify them, sequence diagrams (or sequence graphs; both terms are used) represent a useful modelling tool for thinking about the interactions that occur between objects.

10.6.1 The statechart

Like the *state transition diagram* that was described in Section 9.3, the statechart is concerned with describing the behaviour of a 'system' as a form of finite-state automaton, or finite-state machine. Used at the system level it can be a good way of modelling the behaviour of reactive applications, responding to external events. Used at the object level, it can describe how an object responds to the requests created by its 'response set'.

The statechart was devised by David Harel, and he has observed that it is based upon the more general mathematical concept of the *higraph*

(Harel 1987, Harel 1988). It provides a rather more flexible modelling form than the STD, and in particular, it incorporates the facility for creating a hierarchy of abstractions. It also offers the facility to describe transitions that are orthogonal, in the sense that they can occur completely independently of each other, so making it possible to model transitions that can occur in parallel. And like the STD, it can be used to model the behaviour of a 'problem' (black box) as well as of a 'solution' (white box). The UML has largely adopted the form of statechart devised by Harel (but of course, not quite).

The original paper (1987) provides an excellent tutorial on statecharts and their powers of description. In this, the author uses the functions of a digital watch with two alarms to provide examples. Our examples here are somewhat less dramatic and extensive, but should provide essential ideas about how to use this form.

A *state* is denoted by a box with rounded corners, labelled with the name (identifier) for the state in the top of the box. Hierarchy is represented by encapsulating states within states, and directed arcs are used to represent a transition between states. The arcs are also labelled with a description of the event that triggers that transition, and optionally, with a parenthesised condition. (Conditions are quite common in the real world. For example, it might not be possible to engage a cruise control mechanism in a car—representing a change in its state—while the car is accelerating.)

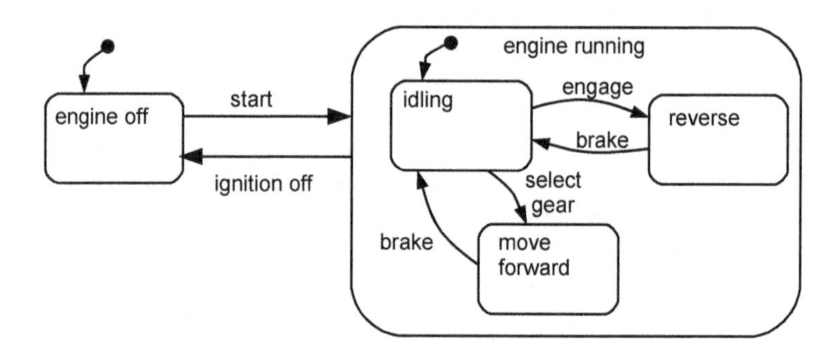

FIGURE 10.13: A simple statechart describing driving a car

Figure 10.13 uses a statechart to provide a very simple state model that describes the use of a car. At the top level it simply has two states that represent the car when the engine is off and when it is running. The default initial state is indicated by the short arrow with a black dot on the end. Transitions between these states are caused by pressing the starter (or waving a card) and by turning the ignition off. There are three sub-states when the engine is running. The initial state has the engine idling and no gear selected. To change to going forward we select a gear, and to stop we brake (OK, there are other ways, but this is a simplified diagram). Engaging reverse we have labelled the transition as 'engage' to emphasise that there is usually only one such gear, and again, stopping involves braking. And of course, there are no transitions between going forwards and reverse (again, this might be possible, but it is usually extremely unwise).

This is a very simple model that wouldn't even be recommended for one's first driving lesson (it doesn't allow for slowing down and accelerating when moving forward as one example), but it is sufficient to show the essential ideas, and particularly the use of hierarchy.

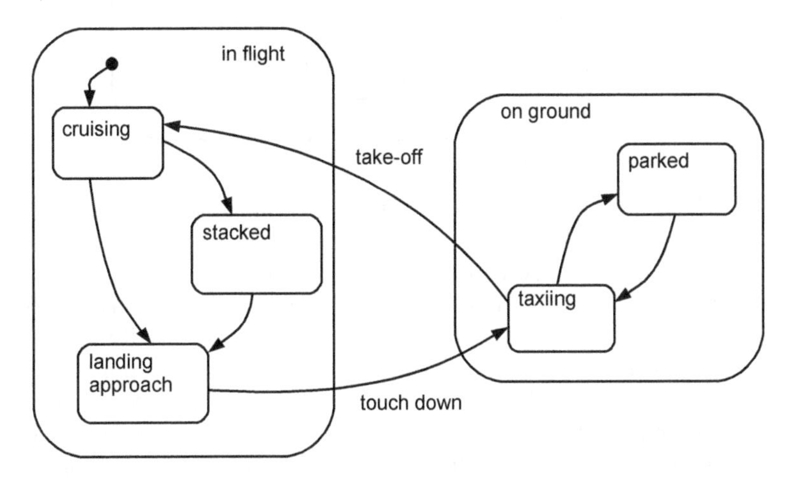

FIGURE 10.14: A simple statechart describing an aircraft in an ATC system

Modelling watches and cars has the advantage that the events that cause transitions between states are directly identifiable in terms of button presses or using a gear lever. Figure 10.14 reworks the example provided in Figure 9.6, describing an air traffic system, to use a statechart formalism. (Some of the labels for the internal transitions have been omitted for clarity.)

When comparing this with the STD used in Figure 9.6, we can see that while the descriptions of state, event and transition are common to both, the STD provides a more detailed description in terms of the associated actions, while the statechart has more refined mechanisms for describing abstraction, defaults and scale. The lack of hierarchy limits the STD to being used to describe the behaviour of individual design elements, whereas the statechart can be used to describe complete systems through a hierarchy of diagrams and states. To some degree the strengths of the two forms are probably complementary, with the STD perhaps being more suited to modelling descriptions of 'problems' and the statechart being better suited to describing detailed 'solutions'.

The remaining major feature of the statechart that should be described here is *orthogonality*. Our example here, shown in Figure 10.15, is rather more abstract in order to highlight the mechanisms involved.

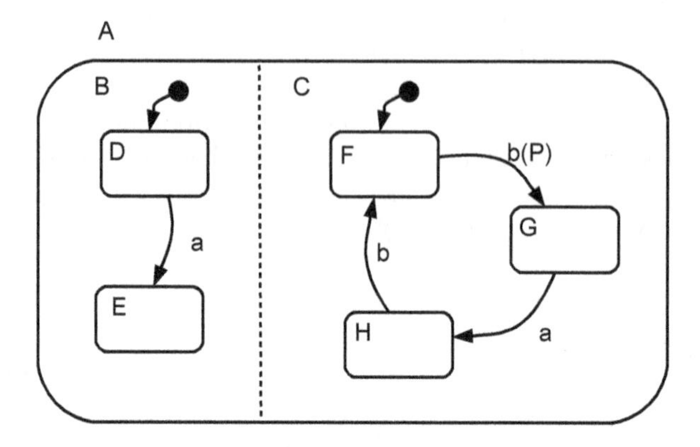

FIGURE 10.15: Describing orthogonality in the statechart formalism

In our example, state 'A' can be described as being a superstate of two orthogonal states 'B' and 'C'. These can in turn be described in terms of further states 'D' to 'G' but the two states involve what are essentially independent groupings. However, events may be common as can be seen from the description of the transitions that occur in response to event 'a'. Also, in describing A, note that we need to identify the default entry to both states 'B' and 'C' (the inner states 'D' and 'F'). There is also a conditional transition between states 'F' and 'G', where an event 'b' will cause a transition only if condition 'P' is satisfied.

The statechart is a powerful modelling tool for thinking about objects, and can also be used for modelling of executable use cases/scenarios (Harel & Gery 1997). Again, it is easily sketched on a whiteboard or on paper.

10.6.2 The message sequence diagram

Sequence diagrams have proved to be particularly useful with object-oriented modelling. This is probably at least in part because they provide a means of modelling the *interactions* between different elements of a system, whether these be objects, actors, remote processes, or client and server. Since essentially they model *collaboration*, in this case object collaboration, they can be used with any architectural style that has loosely coupled elements.

Sequence diagrams can also be useful for modelling the interactions involved in a *use case* (or more correctly, in a specific scenario). Use cases have proved to be a useful tool for modelling object-oriented systems, so this tends to reinforce the value that they have for this type of modelling. The UML (inevitably) incorporates sequence diagrams, describing their form as follows.

> "*A diagram that shows object interactions arranged in time sequence. It shows the objects participating in an interaction and the sequence of messages exchanges*" (Rumbaugh et al. 1999).

A sequence diagram can be viewed as being a dynamic interpretation of a class model. It is concerned with *how* and in what *order* those elements in the class diagram interact, whereas the class diagram simply identifies that they do interact and collaborate. For this chapter at least, they have been categorised with the *behavioural* viewpoint because they essentially deal with interactions between objects (and really don't fit with any other viewpoint either). And like many behavioural forms, it is not hierarchical, and hence doesn't cope particularly well with large-scale patterns of interaction between many elements. And again, like other behavioural forms they can usefully be employed to describe 'problems' as well as 'solutions'.

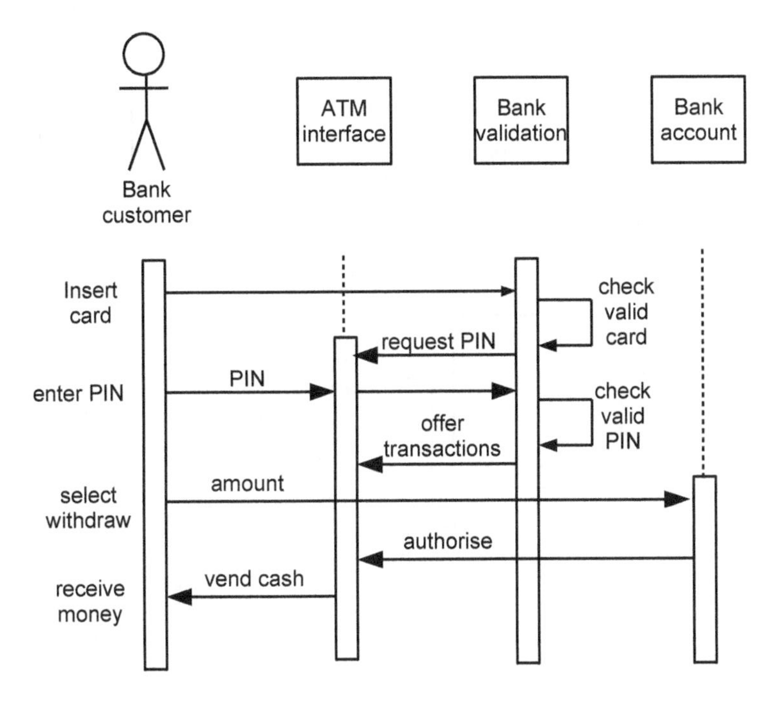

FIGURE 10.16: A simple example of a sequence diagram

The organisation of a sequence diagram is dominated by the use of a *time-line*, which conventionally runs from the top of the page downwards. (Obviously, where there is a need to indicate the lapse of specific time intervals, such a diagram might be drawn to scale on the vertical axis, but generally only the sequential organisation is described. Each processing element, usually a class or object, or possibly some other system 'actor' is allocated a column and messages between them are shown as horizontal lines between the columns.

While the UML provides a very extensive and detailed syntax for modelling with sequence diagrams, it is worth noting that they are relatively easy to sketch and only need limited annotation. Indeed, they can be drawn with a quite minimal set of symbols. Figure 10.16 shows a simple example of a sequence diagram based upon drawing money from a 'hole-in-the-wall' bank machine. This shows a strength of the notation, in that it is easy to discuss whether or not this shows the 'right' distribution of tasks, or issues that have been omitted (such as what happens if the cash dispenser is empty). It also demonstrates a limitation, in that any sequence diagram used to describe a system model or a design model tends to illustrate just one use case/scenario (Ratcliffe & Budgen 2001). (In this role it is known as an 'instance' form of sequence diagram. There is also a 'descriptor' form that can be used for systems analysis and can be used for describing all possible scenarios. Since our main concern here is design, we focus on the instance form in this section.)

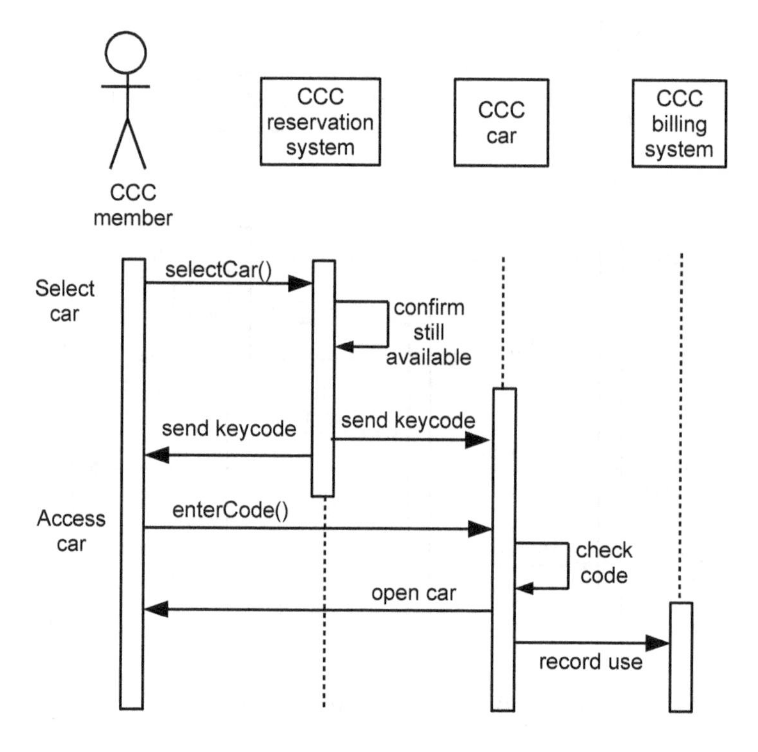

FIGURE 10.17: Designing with a sequence diagram

Figure 10.17 shows a sequence diagram being used in what is more of a design role, and being used to model how a CCC member will perform the operation of selecting and accessing a car. So now the elements at the top of the time lines are classes (or objects) rather than external entities as in our previous example.

This also reflects a wider issue that the objects involved in this sequence must be appropriately related in the corresponding class diagram. Maintaining this form of consistency is a challenge, particularly during more detailed design activities.

10.7 Modelling function: the activity diagram

One viewpoint where the UML does make a useful contribution to the set of notations is the functional viewpoint. UML *activity diagrams* can be used to model the way that a business operates and, like DFDs, can help with analysing the needs of a problem. (There are no really good notations for describing function when using objects. This is perhaps not surprising as unlike many processes, an object often doesn't have a single functional task.)

In particular, an activity diagram can be useful for modelling the type of 'coordinating' situation where a given computation cannot proceed until certain conditions are met (for example, new data is available *and* a prior computation has completed). The diagram does model states, but these now represent the performance of actions, and the focus of interest is upon what triggers transitions between the states. This emphasis upon *action* is why it has been categorised as functional here. However, rather as with sequence diagrams, an activity diagram can be viewed as providing a mix of functional and behavioural aspects in its description.

Some key notational elements are as follows.

- The *activity*, which is a task performed by the application and is shown as a named box with rounded sides.
- A *state*, which can be viewed as being an activity where nothing happens.
- A *transition*, where work flows between activities, shown as an unlabelled arrow. The lack of a label is because, unlike the case of a statechart, the transitions arise from the actions of the activities themselves, not because of external events.
- A *decision*, represented by a diamond, where a workflow divides between possible branches. When used, it is then necessary to label the transitions to indicate which condition is employed for a particular route.
- The *synchronisation bar* is a thick horizontal line that represents the coordination of the activities. When all of the transitions into a bar are complete (the coordinating bit) then the outward transitions will be 'fired'.

Entry and exit from a diagram are shown respectively by a short arrow with a filled dot on its tail, and a filled dot with a circle around it. There can only

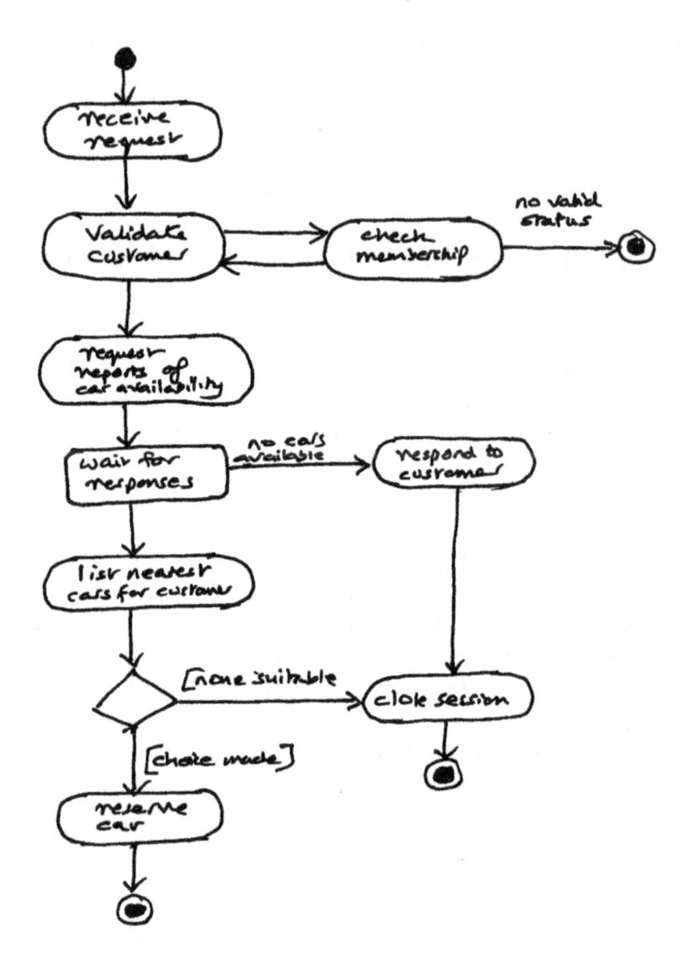

FIGURE 10.18: An activity diagram describing booking a car

be one entry point, but there may be many exit points. This is illustrated in the sketch of the CCC booking process shown in Figure 10.18.

Figure 10.19 uses the bank teller machine example to illustrate the use of synchronisation bars. One shows division of transitions (where multiple actions occur after a coordinating action, termed a *fork*), the second where an operation can only proceed after two transitions have completed (a *join*).

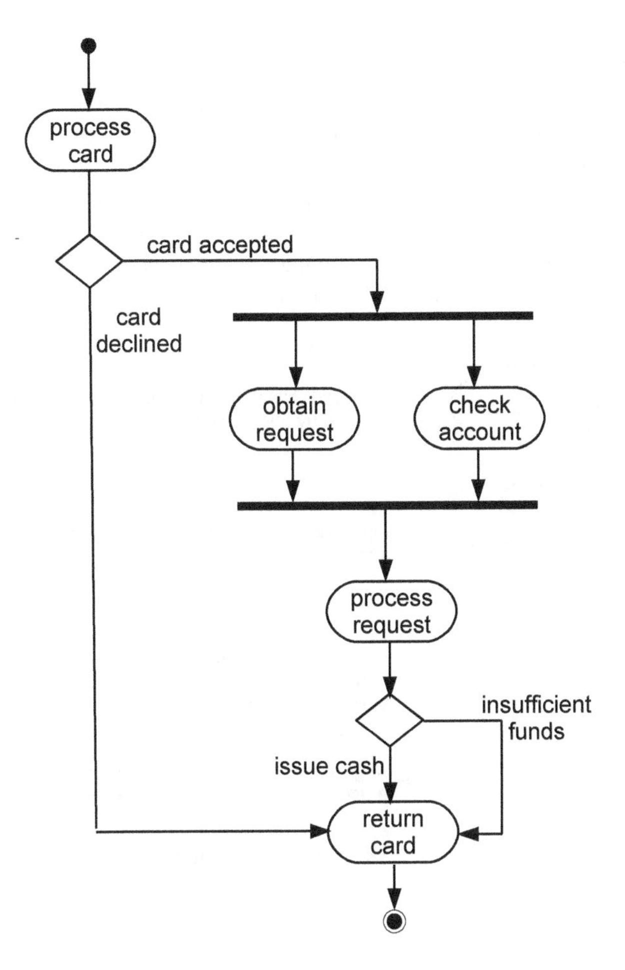

FIGURE 10.19: An activity diagram using synchronisation bars

10.8 Use cases

This is a good point to discuss something that has been mentioned previously in this chapter, namely the idea of a *use case*.

The idea of the *scenario* describing a particular way that an application is used is one that has been employed informally over many years as a convenient description of, and way to think about, system behaviour. Ivar Jacobson made a valuable contribution to design modelling when he generalised this idea into that of a *use case*, formalising it as a tool for thinking about software design (Jacobson, Christerson, Jonsson & Overgaard 1992). Not surprisingly, his involvement in the formulation of the UML meant that use cases are a key element of this too.

We can draw an analogy between the relationship between a use case and a scenario and that of a program and a process executing on a computer. A program describes the rules for performing a set of possible actions, and a process comprises a particular instantiation of that program. Similarly, a *use case* represents a set of possible interactions between an application and other *actors* (where these can be devices, people, other pieces of software etc.), while a *scenario* will describe a particular sequence of these interactions.

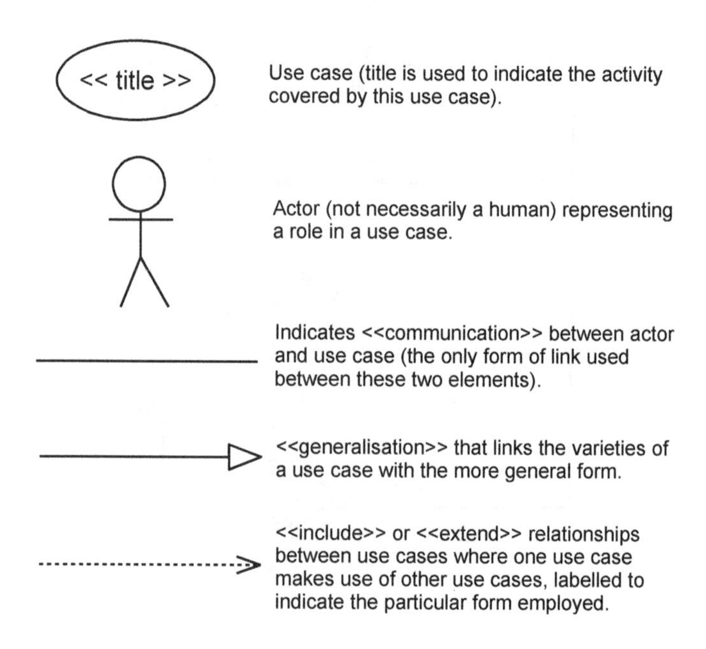

FIGURE 10.20: Elements of a UML use case diagram

A major attraction of employing use cases is the ability to document a set of user interactions with the application, and then to use the associated message sequences for a given system configuration to guide the design of an object model that will implement that application behaviour. Using a common abstraction to describe both behaviour and object model can help with the design modelling process.

The UML *use case diagram* expresses the idea of a use case at a fairly high level of abstraction. Figure 10.20 shows the basic elements of a use case diagram. The use case itself is shown as an oval, and represents "a logical description of a slice of system functionality" (Rumbaugh et al. 1999). Actors are (perhaps confusingly at times) usually represented by stick figures. Finally, there are four kinds of relationship shown in the figure, although two of these share a single representational form and hence need to be labelled with a stereotype in order to distinguish them. As we should note, the use case diagram is easy to sketch, and made even easier when designing because for most purposes, we only need the use case, actor and communication symbols.

Like other descriptive forms that are used with objects, the use case diagram is difficult to classify using the viewpoints model. The use case is concerned with the interactions that occur between a system and its environment, and hence can have a *behavioural* aspect. But they are also concerned with the tasks that an application performs, and so have a *functional* aspect in their role. (Since we can sometimes describe scenarios and use cases using message sequence diagrams, this mix is to be expected.)

Figure 10.21 shows a simple use case diagram for part of the CCC, again relating to the car reservation process.

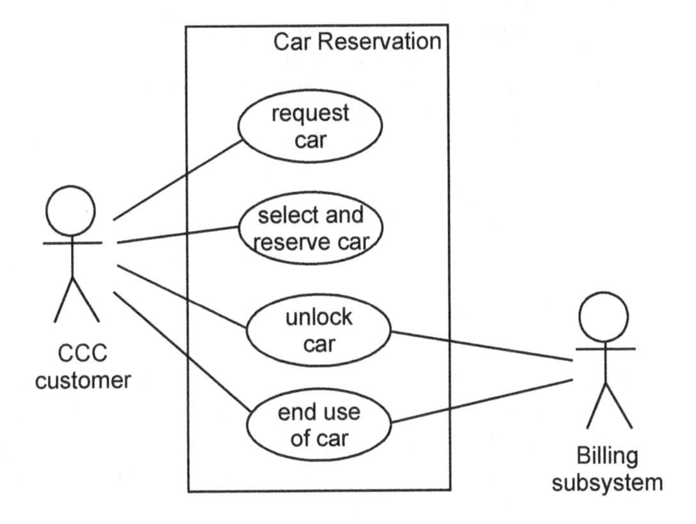

FIGURE 10.21: Example of a UML use case diagram

In this model, the process of reserving a car involves four use cases, one to handle the initial request, another to negotiate the choice of a car, the third concerned with the user opening the car and finally depositing the car at the end of the session. The customer is an actor in all of these, while the billing subsystem is only an actor participating in the last two (when the customer actually claims and uses the car).

The UML use case diagram itself is largely concerned with the identification of use cases and the set of actors acting as participants in these. The means of creating a detailed specification of a use case is largely a matter of choice. It might involve the use of sequence diagrams, and possibly of some form of textual specification too. The latter might identify such aspects as the participating actors, pre-conditions for the use case to occur, details of actions involved and then post-conditions when the use case ends. Figure 10.22 provides a simple example for one of the use cases in Figure 10.21.

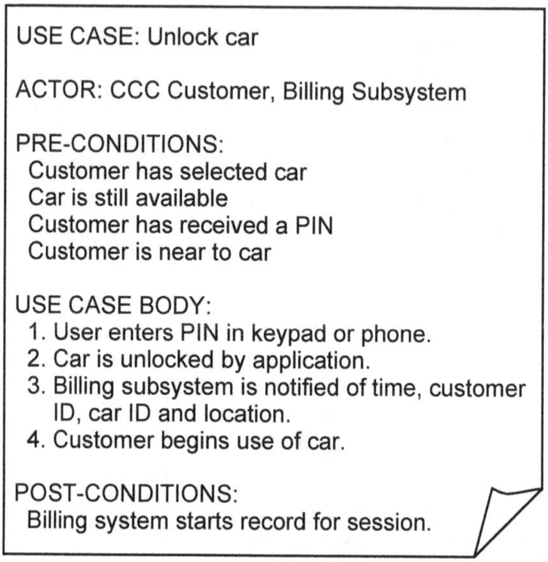

USE CASE: Unlock car

ACTOR: CCC Customer, Billing Subsystem

PRE-CONDITIONS:
Customer has selected car
Car is still available
Customer has received a PIN
Customer is near to car

USE CASE BODY:
1. User enters PIN in keypad or phone.
2. Car is unlocked by application.
3. Billing subsystem is notified of time, customer ID, car ID and location.
4. Customer begins use of car.

POST-CONDITIONS:
Billing system starts record for session.

FIGURE 10.22: Example of a UML use case

The *Unified Process* (UP), described in Chapter 13, inevitably employs use cases, and they have been widely employed formally and informally. Use cases can also be related to the concept of the *user story* as employed in agile software development (Jacobson, Spence & Kerr 2016). They are not particularly dependent upon the use of an object-oriented architecture either, although they have made a particular contribution to applications of that form.

10.9 Empirical knowledge about modelling objects and classes

This is a large chapter, and so this section has been organised around three important aspects of object-orientation that are described here. The first is the *object model* itself; the second is the different notations used for the UML; and the third relates to metrics used with objects.

10.9.1 The object model

While classes and objects have a long history in software development, stemming back to the 1960s, and have been employed widely for that purpose, there is surprisingly little in the way of empirical studies related to designing applications using objects. A systematic mapping study examining this (Bailey, Budgen, Turner, Kitchenham, Brereton & Linkman 2007) found only 138 studies, with nearly half of these being concerned with metrics, 10 with the use of design patterns, and 19 with comparing the use of OO with non-OO forms. The most widely-used form was the laboratory experiment (arguably not the most effective tool for studying long-term design issues) with observational studies and case studies making up the form of empirical study for just under half of the papers found.

What that mapping study did not include was what we term *experience papers* and material from the grey literature. It may well be that these are used very widely to record design experiences in an informal manner.

Perhaps we should not be that surprised at the relatively small number of formal studies. There are even fewer studies looking at earlier forms of software architecture such as call-and-return, and the nature of design is such that making simple comparisons (such as call-and-return versus object-oriented) is a fairly pointless exercise. Questions that are more useful are going to be those relating to such issues as the relative ease of learning how to use different modelling forms (such as the study by (Vessey & Conger 1994), and guidance about how to use the different modelling approaches.

The systematic review by Tiwari & Gupta (2015) examines the evidence about the different roles performed by use cases. They found that use case specifications were typically employed for two perspectives. The first was for documenting functional requirements, while the second was for generating the lower-level software artifacts, particularly through model transformation.

10.9.2 Object modelling notations

Essentially, the only widely documented source of object modelling notations is the UML. In addition to the analysis of visual notations by Moody (2009) which discusses many of the UML notations, there have been a number of studies looking at particular notations. The systematic review reported in (Budgen, Burn, Brereton, Kitchenham & Pretorius 2011) identified 49 studies, with metrics and comprehension being the two topics studied most extensively. Most studies were laboratory studies, with relatively few 'field' studies.

The subsequent survey by Petre (2013) also emphasised that the UML was not widely used in practice, although it was sometimes used for documentation. Of course, while the UML was not used formally, this does not rule out the possibility that its forms did influence the informal notations used by designers.

10.9.3 Object-oriented metrics

The systematic review by Radjenović et al. (2013) looked at the value of different metrics for software fault prediction. It found that the Chidamber & Kemerer metrics were the most successful of the available OO metrics, with CBO, WMC and RFC being the most successful, while DIT and NOC were unreliable.

We might note here that from a measurement-theoretic perspective, LCOM is considered to be 'theoretically invalid' (Litz & Montazeri 1996), although from our perspective, it does identify a design property that may be important for objects. Kitchenham (2010) also observes that CBO is 'flawed' because it treats "forward and backward links as equivalent" arguing that these may not present equal difficulty of understanding, and hence have different influence upon likely faults. Again, from a design perspective, our main concern is the existence of such links, rather than their role.

Key take-home points about modelling objects and classes

Modelling the attributes of objects and classes uses a range of different forms to address the main viewpoints for each type of design element.

Objects are complex. The structures and properties of objects (and classes) are much more complex than those of processes, necessitating the use of more viewpoints together with the need to model a range of relationships as well as the elements.

Classes and objects. Both classes and objects are important design elements when developing design models—and there is no 'right' abstraction to employ for describing their properties in all circumstances. They can also be difficult to identify when creating the object model.

Relationships can have many forms. This aspect of object modelling can be challenging since there are many forms of relationship that can occur between objects, with properties that may need to be represented in different ways.

Learning to model objects is challenging. Studies suggest that object-oriented modelling is harder to learn than the use of simpler forms such as processes.

Modelling objects needs a mix of notations. The different characteristics require a mix of forms, and these also need to be kept consistent (a potentially useful role for tools).

Object notations make poor use of visual separation. Popular forms of diagrams, such as most of those used in the UML, tend to use a limited set of shapes and symbols, and make excessive use of shapes such as oblongs. The range of possible relationships adds to this, as it is even harder to devise distinct visual forms for these.

Chapter 11

Modelling Software Components and Services

This chapter looks at two quite different approaches to the idea of reusing software that has been developed by others. The concept of the *software component* has been around for many years—indeed, the concept of reusable blocks of software was put forward at the inaugural Software Engineering conference in 1968 by Doug McIlroy. Building on this concept, *Component-Based Software Engineering* (CBSE) has had the goal of enabling a software developer to be able to make use of pre-existing reusable 'chunks' of software in order to create (or 'compose') an application. In many ways the *service paradigm* represents an evolution of this idea. While it largely aims to do the same thing, the means of realising it is very different. Here the software itself is provided by third parties, and is hosted and executed elsewhere (typically in the *cloud*) rather than being embedded within the application, enabling easy substitution of different providers.

Each of these technologies poses challenges regarding both about how to design the reusable elements themselves as well as how to design an application to make use of existing resources. For the moment though, we concentrate on the forms those resources can take and how to model them, leaving the question of how to design applications around them until a later chapter. However, since *reuse* is so important for both, we begin by discussing its role and influence.

11.1 Reuse

For other domains in which design performs an important role, the idea of *reuse* is fairly well established and takes a range of forms. It is sometimes associated with the abstract reuse of design ideas and experiences, rather than of physical elements, and we will look at this aspect in Chapter 15 when we examine how the concept of a *pattern* can be employed.

Where the manufacturing cycle is important, then it is also associated with the interchangeability of physical *components*, not least because creating 'product lines' can help reduce the cost of the individual product[1]. Mechanical engineering in its many forms makes extensive use of components, as do the electrical engineering and construction industries. In all of these, the use of standardised units helps with manufacture and maintenance. Their use also changes the nature of the design process by introducing the notion of *composition*.

Organisational and other processes themselves can be reused too. We can see this in the way that production lines in factories can be re-organised and re-purposed to create new products.

Successful reuse of pre-existing elements is commonly based upon two important characteristics.

- The first is that the role of a reusable component is associated with some well-defined *functionality*. Most physical components have very well-defined roles (switch, pump, door handle, starter motor, etc.) making it possible for the designer to select suitable items from a catalogue.

- Secondly, the components themselves usually have well-defined *interfaces*, enabling ready composition with other components that meet the same standard, as well as easy substitution of one manufacturer's component with one from another maker. This can also help with identification of suitable ones in a catalogue. In manufacturing this can mean that there may be several manufacturers who are able to provide a particular component ('second sourcing'), providing users with confidence that they will be able to obtain adequate supplies.

These are not accidental properties, but rather, they are ones that arise from economic factors. These factors include the need to minimise the cost of fabricating an item; pressure to maintain a marketplace position; and the end-user's need to protect their supply chain (here the end-user will be the manufacturer who is making use of a given component in their own product). The adoption

[1]One of the earliest examples of a product line is probably when Sir Marc Brunel (father of Isambard Brunel) established his block-making machinery in Portsmouth dockyard around 1806. This introduced an important element of standardisation in the fitting out of sailing ships, which used large numbers of blocks as part of their rigging.

of interface standards may be motivated in a number of ways: one manufacturer may be so dominant that their interface becomes widely accepted; an industry (or professional body) may take on the role of defining the necessary standards (as has happened widely with electrical connectors); or a group of manufacturers may come together to agree on a common standard.

Reuse can be facilitated by the availability of catalogues of components. However, the way that such catalogues are used during the design process appears to be less well understood. Pugh (1991) provides an example of this in the domain of electronic engineering, in quoting from Cooke (1984).

> "*The designer of electronic-based products has available an enormous range of basic building blocks of high or still increasing complexity. What he often does not have is either training or well-established conceptual tools for working out levels of sophistication.*"

Similarly, in Pahl & Beitz (1996), a textbook that is widely cited as a repository of engineering design knowledge, the discussion of 'design catalogues' is almost entirely confined to consideration of how such a catalogue will be *constructed* rather than how it might be *used*.

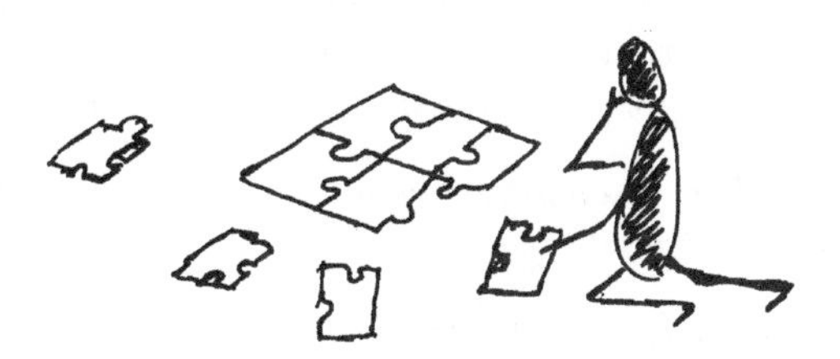

If we turn our attention to the reuse of software, we find that, despite the lack of a manufacturing phase, reuse of software components has had a long history of success in some areas. An often-cited example is the set of mathematical and statistical functions provided in the NAG (Numerical Algorithms Group) library that was first released in 1971. The set of functions and the programming languages supported has gradually expanded over the years, and the routines it provides do of course meet our two criteria above, having well-defined functionality and well-defined interfaces. Other examples include user interface packages such as *tk* and tools such as *MATLAB*® that provide an environment of 'components'. *Component models* such as CORBA (Common Object Request Broker Architecture) have assisted further by defining interface standards to enable local and distributed components, possibly written using different programming languages, to work together.

However, because software is so easily adapted and modified, it would also be true to say that examples of wide reuse, in the sense of embodying one software product within another, are relatively rare. Where they do occur then there may be special factors that constrain this freedom. (Network protocols provide a good example of this—there is little point in producing unique network protocols if you want your software to work with other applications.)

Indeed, although the benefits of reuse are often extolled by software engineers (and vendors of software), for example, when considering the benefits of an architectural form such as the object model, attempts to incorporate it more fully into software development practices have not been widely successful. In the rest of this chapter we describe two approaches to software development that focus on reuse, and examine the issues associated with their creation and use.

11.2 Modelling software components

The idea of *component-based software engineering* (CBSE) began to attract greater attention in the 1990s, and in some ways the concept of a *component* formed an evolution of the idea of the *object.* And, as with objects, ideas about precisely what constitutes a component could be difficult to pin down. Indeed, when writing an introduction to a special journal section on CBSE, Brown & Wallnau (1998) observed (slightly tongues-in-cheeks), that:

> "*CBSE is a coherent engineering practice, but we haven't fully identified just what it is*".

Note though, the use of the word 'practice', as it could certainly be argued that one of the motivations of research into CBSE was to seek something that could be more tractably 'engineered' and composed together than objects.

One of the challenges of codifying CBSE was to determine just what a component is. An early definition from Brown & Short (1997) very concisely described a component as:

> "*an independently deliverable set of reusable services*".

Note too that this definition made no explicit assumptions about architectural style—so that a component could be realised as anything from an object to an operating system. The emphasis here is upon *reuse* (needing a clearly specified interface) and upon the idea of *independent delivery.* This second aspect is important since it implies that a component should not be aware of its context, especially in terms of any embedded dependencies or the expectation of the presence of some shared resource. There is also the implication that a component needs to be integrated with other components in order to provide the overall system functionality. In other words, it is a part, rather than a whole.

The pioneering text on software components by Szyperski (1998) provided a rather more extensive definition of what a component was considered to be.

> "*a unit of composition with contractually specified interfaces and explicit context dependencies only. A software component can be deployed independently and is subject to third-party composition.*"

While addressing the same properties as the more abstract definition provided by Brown and Short, this added the concept of *black box* reuse, required if it was to be used by third parties.

A slightly more evolved definition was that provided in Heineman & Councill (2001), where a software component was defined as:

> "*a software element that conforms to a component model and can be independently deployed and composed without modification according to a composition standard*"

This definition is interesting because it separates out two further supporting concepts, which are those of:

- the *component model* that incorporates 'specific interaction and composition standards'; and
- the *composition standard*, that 'defines how components can be composed' to create a larger structure.

Both of these are important concepts in terms of considering how we can design with components, and also how components themselves should be designed in order to be usable. Indirectly, they also raise the question of architectural style, and the extent to which the process of composition may need to be confined to applications using an architectural style that conforms to the 'standard' being used.

None of these definitions are really contradictory. Where they differ is in the level of abstraction required to express the ideas, and which of the many aspects of a component they emphasise. It is worth noting that there may also be business factors that influence the component model and its acceptance (Brereton & Budgen 2000) (a more business-focused view of CBSE is also presented in Brown (2000)).

And as we will see when addressing the second topic of this chapter, the evolution of these definitions has distinct parallels with the emergence of the idea of a *software service*.

11.2.1 Component characteristics

Of course, not only have ideas about the nature of a component evolved since the concept was first envisaged in 1968, the forms of components have too. As Crnkovic, Stafford & Szyperski (2011) observe:

> "*Starting as elements of source code, such as routines, procedures, methods or objects, they transformed to architectural units and ready-to-execute blocks that are dynamically plugged into the running systems.*"

In discussing the idea of *reuse*, the two characteristics that were identified as being essential for a component of any form were:

- well-defined functionality
- well-defined interfaces

A further characteristic, which is probably implicit for non-software forms of component, but which needs to be explicitly stated for software components (because of the nature of software) is that of *independence.* This avoidance of any context-specific dependencies follows on from the ideas above. In practice of course, some dependencies are difficult to avoid, particularly where these related to issues such as architectural style, but where these do exist, they need to be made fully explicit. Figure 11.1 illustrates this set of characteristics.

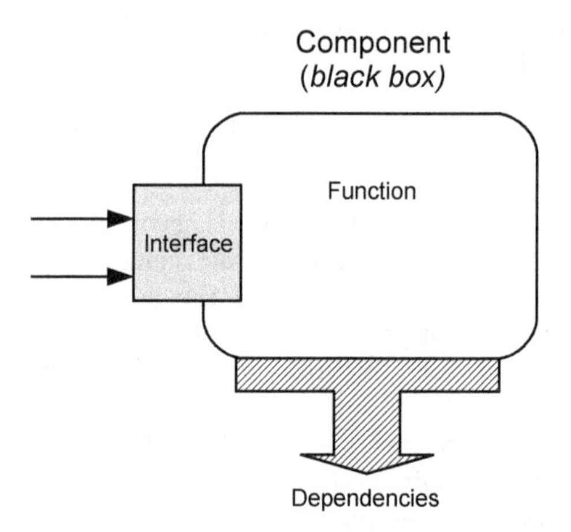

FIGURE 11.1: Characteristics of a software component

If we look at other domains, we might note that there are often implicit context-specific dependencies. For example, the starter motor for a car will depend upon the availability of an electrical supply with a specific voltage and power rating. However, software dependencies can be more subtle, and so to ensure that a software component can be treated as a black box as far as is reasonably possible, such dependencies need to be made explicit in the specification. (As an example of the need for this, and of the difficulty of ensuring it is comprehensively achieved, see the analysis of the $500 million failure of the Ariane 5 rocket in 1996 that is provided in Jézéquel & Meyer (1997).)

We might note here that the UML *component diagram* is a very specific interpretation of the concept and of little use in a CBSE context. Indeed, perhaps because the concept of a component lacks any association with a specific architectural style, there are no diagrammatical forms that have been widely used for modelling CBSE implementations.

However, that said, many aspects of component use can be modelled using existing notations. In particular, issues related to *functional* and *behavioural* modelling can often be modelled using the forms commonly used with objects, such as activity diagrams, statecharts and sequence diagrams. What is not so readily modelled is the constructional viewpoint, although for many purposes, components can be modelled by adapting existing forms of class diagram.

11.2.2 Component frameworks

Commonality of architectural style does not guarantee that components of different origins can be easily integrated to create an application. This issue, termed *architectural mismatch* was first identified in the context of seeking to compose an application by making use of existing objects (Garlan, Allen & Ockerbloom 1995). Attempts to create new applications from objects that had been created by different sources revealed that these made different, and incompatible, assumptions about the way that the object model was organised. When the original analysis was revisited in (Garlan, Allan & Ockerbloom 2009), the authors concluded that the reuse of components still formed a challenge, and indeed, that while the expanded computing landscape had reduced some of the earlier problems, it also introduced new ones.

What sort of assumptions created the problems? In the original (1995) study there were four general categories of assumptions that could result in mismatch. These were:

- the nature of the (other) components, including the control model;
- the nature of the connectors and the protocols used;
- the global architectural structure; and
- the construction process.

There were also three aspects regarding the ways that components interact where assumptions could produce mismatch: the infrastructure; the application itself; and interactions between peer components. None of these were issues that are strictly matters of design, although they might constrain design choices.

The later study identified some changes that had altered the situation; greater use of reusable infrastructure; use of standard interfaces such as web browsers; extensive use of open source software; and more sophisticated development environments. They also noted the emergence of architecturally specialised design domains. However, this more distributed context introduced such issues as *trust* between components; *dynamic* reconfiguration of systems; system evolution; and technical debt in the form of being locked into a specific architecture.

The idea of the *framework* as a means of imposing a common component model has formed an important element in the evolution of CBSE and of components themselves. It has also helped to address many of the issues identified above. While early ideas, such as those illustrated in Figure 11.1 assumed that components would be integrated to form monolithic software applications, the use of frameworks has allowed for more loosely coupled and distributed use of components, while also providing the necessary protocols for interaction between components. Figure 11.2 illustrates this evolution.

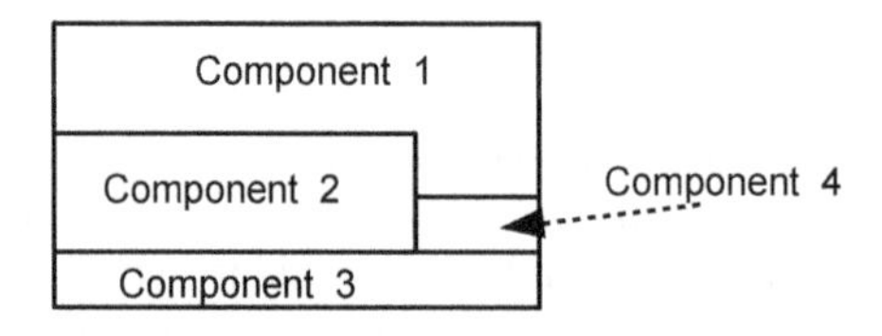

a) Monolithic integration

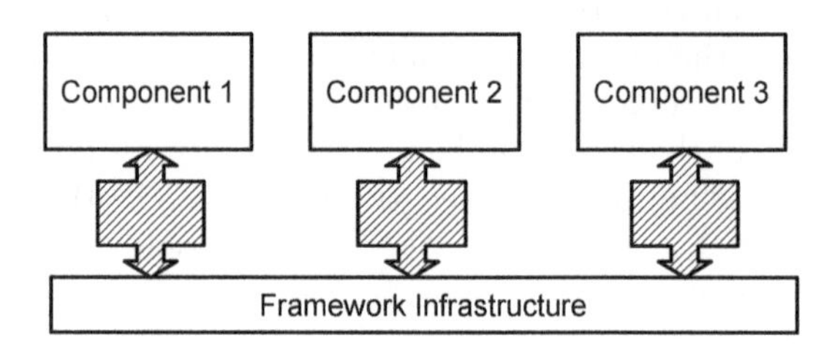

b) Integration via a Framework

FIGURE 11.2: From monolithic construction to frameworks

So, what is a *component framework*? It can be summarised as being a set of formal definitions for the interfaces between components, where the definitions establish a set of protocols through which components will interact and cooperate within the given framework. In turn, these protocols may include 'naming conventions' used for the methods in an object that provide particular functions, the requirement to provide particular functions, and the operational procedures required for a component to declare itself available for use, and for others to request its services.

There are many frameworks (of course). We have already mentioned CORBA, which is maintained by the OMG (Object Management Group) that we encountered in the last chapter in the context of UML. CORBA has mapping to many implementation languages and the 'client' of an object can access that object only through the protocols of its published interface. There are a number of frameworks provided by commercial vendors, such as Microsoft's .NET and of course the open source frameworks, such as Apache Spark. There are also many frameworks aimed particularly at Web applications, with J2EE providing an important Java-based example.

For the purposes of this chapter, the details of particular frameworks are of less importance, what matters is that they exist—and that the choice of a framework has therefore become one of the design activities. And like other design activities, this is one that can involve complex trade-offs between many different factors—and inevitably, the choice creates a long-term technical debt.

11.2.3 Designing components

The design of the actual components can be considered as having two goals. The first of these is to achieve the three general component characteristics identified earlier, in order to ensure that a component can be "deployed independently". These require that a component:

- possesses well-defined functionality;
- provides a set of well-defined interfaces;
- clearly specifies any explicit dependencies.

And over and above that is the second goal, of conforming to the standards of a particular component framework. This aspect is likely to be supported by a range of tools (depending on the choice of framework), even down to the naming convention for any identifiers that are to be used for externally-visible methods.

Returning to the question of *functionality* (a major motivation for creating components), designing components to be reusable does imply that the designer should strive to make the component as general as is reasonably possible. This is not an argument to use an excess of 'bells & whistles' or a 'swiss army knife' approach, indeed quite the opposite. Making a component do one job and do it well was an important philosophical underpinning for the Unix operating system when it was developed in the 1970s. Using simple tools that connected together readily via standard mechanisms helped users create their own applications with only limited effort (or even programming knowledge). And as Unix continues to underpin much of our computing infrastructure, we can reasonably assume this was a good principle to adopt elsewhere.

Experts prefer solutions that they know work. PvdH #13

The paper by Crnkovic & Larsson (2002) provides an interesting case study on component use. The authors examine the evolution of two components within the application being studied (a control system package used for industrial applications). While not specifically concerned with component development, this still examines how key component characteristics affect the two particular aspects being studied: evolution of components across platforms and replacement of proprietary components with industry-standard ones. And, although the two components have relatively low-level roles (one can be considered as middleware, while the other is a class library), the general experiences may well have wider generality.

Some of the particular issues that the case study identifies from the viewpoint of designing individual components are as follows.

- The benefit of using larger components that are easy to reuse, in preference to smaller ones that could be developed in-house.
- The high cost of maintaining backward compatibility with early installations (this arose largely because the application being studied was a 'product line' so creating additional technical debt).
- The need to separate out those features of a component that depend upon platform interactions in order to assist with ease of change.

It is worth noting here that components can be relatively costly to create. In Szyperski (1998) it is suggested that developing a reusable component could require 3–4 times the resources needed for a 'non-component' implementation. However, since some of these additional costs arise from external factors such as evolutionary changes to the component framework, it is difficult to predict the additional overhead involved in using a component. However, it is worth noting that unless there really is good scope to make reuse of components, or an application can be largely created using existing components with only a few additional specialised ones being built, developing components may not make economic sense. (A good example of where extensive use is made of existing components is the common practice with web applications of making use of component frameworks to build a user interface.)

11.2.4 COTS

COTS stands for "Commercial Off The Shelf". Use of COTS components can be regarded as representing an extreme, in the sense that the end user (designer) or integrator has no control whatever over their form and properties, and no knowledge about their workings. So a COTS element is one that is quite expressly black box in its nature.

However, in a software context, this view is probably something of an extreme position. As Carney & Long (2000) point out, what is usually termed COTS software spans something of a range of forms, some of which are capable of some degree of parameterisation, and with the extent of any modification often depending upon the financial and political 'clout' of the customer.

The main issue with the use of COTS is really concerned with the integration of such components, and so we leave further discussion of the use of such components for Chapter 16.

11.3 Modelling software services

As described in the preceding section, and illustrated in Figure 11.2, concepts about how software components could be composed evolved with developments in technology. Early ideas about CBSE were centred upon using tightly coupled integration to form a monolithic application image that would execute on a local computer. With time, this evolved to employing much more loosely coupled forms, using frameworks for interconnection, with the components possibly being distributed across a wide network. And although there was always some expectation that components could be provided by third parties, with the NAG library being a good example, the use of more loosely coupled environments introduced greater emphasis upon related issues of trust and dynamic reconfiguration.

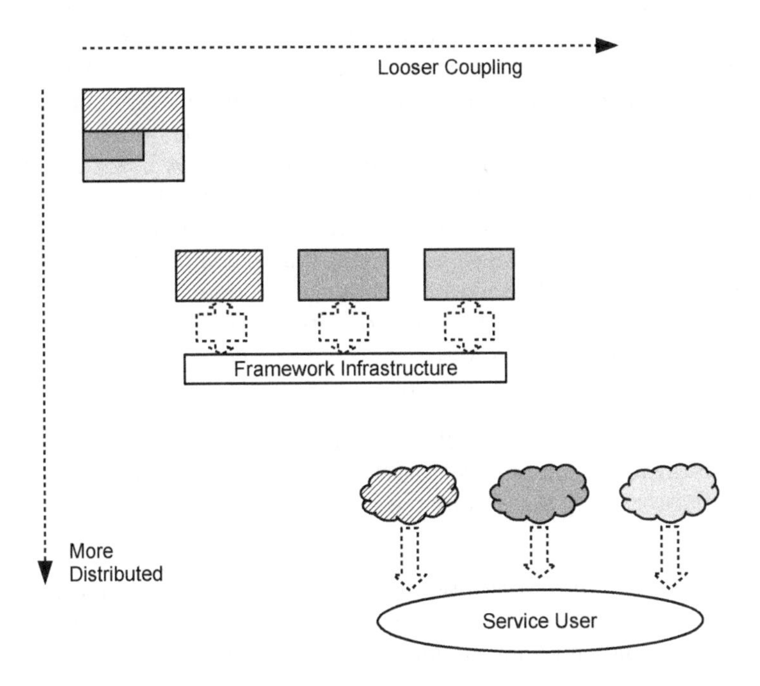

FIGURE 11.3: The evolution from components to services

The concept of a *software service* can be viewed as representing the next step in that evolution of reusable software components. Indeed, one way of thinking about services is as remote components that are very loosely coupled, so that the application making use of them does not need to know anything about how they are realised. This progression is illustrated in Figure 11.3. Here the service sources are shown as miniature 'clouds', and indeed, the idea of software services has been an important underpinning for the concept of a *cloud*.

Service technologies, in one form and another, emerged around the millennium, and the different practices of using services to create applications have resulted in a variety of descriptive terms such as *Software as a Service*, *Service-Oriented Architecture* (SOA), *web services*, as well as a large range of acronyms that describe the protocols used to support the various activities involved (Turner, Budgen & Brereton 2003). Services are usually accessed and delivered across the web, often by third-party providers.

Service models focus upon separating the *possession* and *ownership* of software from its *use*. By delivering software's functionality as a set of distributed services that can if necessary be configured and bound at the time of delivery (*ultra-late binding*), provides a highly flexible strategy. Of course, the end user still needs to have some form of application that incorporates a set of business rules, but this can be fairly minimalist and only needs to specify what services are needed, not how they are to be provided. Figure 11.4 shows two forms of service model. Model a) is one where this is organised on a *supply-led* basis, using a pre-determined range of services from a remote provider (a step on from distributed components, but still using a fixed profile). Model b) shows how this can be organised on a *demand-led* basis, involving an additional integration layer.

Basic service description, discovery, delivery and composition are largely organised using XML-based protocols.

SOAP (Simple Object Access Protocol) provides a message format for communicating with, and invoking, web-based services. (REST is also used, but historically, SOAP was the main protocol employed with web services.)

WSDL (Web Services Description Language) is used to describe a service in terms of its acceptable data types, methods, message format, transport protocol and end-point uniform resource identifier (URI).

While these are largely sufficient for supply-led configuration, the additional elements needed for managing demand-led service use require a set of additional protocols to scope out the additional information required.

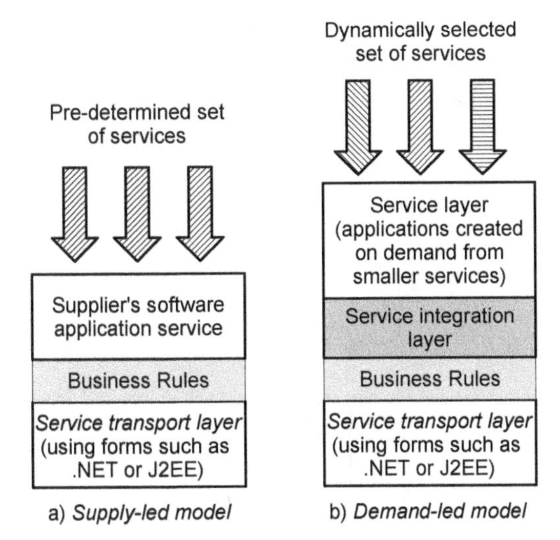

FIGURE 11.4: Supply-led and demand-led service models

Figure 11.5 shows the basic mechanics of service use. The *service consumer* makes a request for a service. This may be a request for a specific service from a particular provider, but if not, the request goes to some element that provides for dynamic composition or orchestration of services (as in the diagram). This then negotiates with one or more possible providers, depending on the business rules that are specified, and then selects a particular *service provider*, which is then bound to the service user for this transaction. The actual service is then provided through the protocols embodied in that particular *service description*.

This is of course a fairly simple single transaction, and providing for a user's need may well require that several services are composed together in order to perform a specific task. Note also that the same symbol has been used for both the service consumer and the service provider to emphasise that a consumer can itself also be a provider.

Modelling both the realisation of a service, and also its use is clearly a fairly challenging issue. There are no specific service modelling notations, and like components in general, most modelling appears to be performed by using existing notations. Those provided by the UML can be useful (with the benefit of having drawing tools available if required), as are DFDs (Anjum & Budgen 2017).

A particularly distinctive element in the process of service composition is the presence of the *business rules*. Although these are normally a factor to be considered when designing applications, they are usually implicitly embodied in the design process (essentially forming an element of the design specification). However, when using software services, the business rules need to be

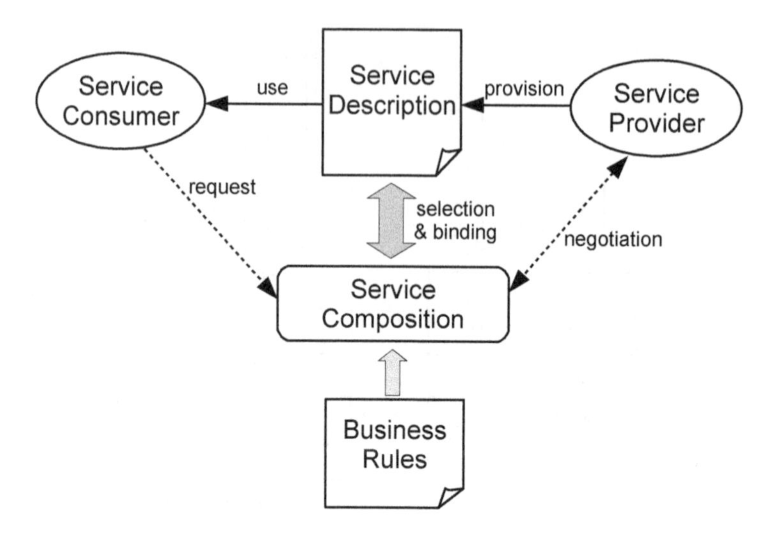

FIGURE 11.5: The elements of service use

made explicit, since ensuring that the application conforms to these needs to be a part of the dynamic composition process. Modelling these rules can add a further challenging element when using a service model.

11.4 Empirical knowledge about modelling components and services

The corpus of empirical studies related to both components and services is limited. As with objects, attempting to perform meaningful empirical studies during the emergent period for a technology, when the element being studied is not particularly well-defined is rather challenging. So, not surprisingly, the available primary studies tend to be a mix of case studies, surveys and experience reports.

11.4.1 Empirical knowledge about components

The systematic mapping study by Vale, Crnkovic, de Almeida, da Mota Silviera Neto, Cavalcanti & de Lemos Meira (2016) looks at studies related to CBSE over the period 1984-2012. The authors analysed some 1231 primary studies and examined the profile of research that these revealed for that period. Not surprisingly perhaps, they observed a decline in publication rates in the 2010s, since as they noted "CBSE has been integrated into other approaches, such as service-oriented development, software product lines, and model-based engineering".

They noted that the main objectives for adopting CBSE tended to be non-functional ones such as increasing productivity, saving on cost, improving quality etc. They noted that there were "few studies that measure the effort involved in building for reuse" and suggest that industry experience is that "building reusable components on average requires three to five times more effort than building the same function but not building for reuse" (similar to Szyperski's estimates).

While the literature identified more than 50 component models, with the four most frequently addressed including CORBA. They also identified a wide range of domains in which CBSE has been applied and observed that the provision of support for these "explains the large number of component models", with these differing chiefly in term of being realised to meet the needs of a domain, rather than in terms of basic principles.

11.4.2 Empirical knowledge about services

Few systematic reviews on the use of software services are available, and those that are, tend to be looking at rather specific research issues rather than at questions about the effectiveness of services.

Key take-home points about modelling components and services

Modelling the attributes of software components and services uses a range of different forms, but to date modelling has tended to use existing notations rather than develop any that are specific to these technologies..

Reuse. This forms a major motivation for the adoption of components, whether close or loosely coupled.

Component characteristics. Components are characterised by having well-defined functionality, accessed through well-defined interfaces, and with any dependencies being made explicit. Ideally, a component has very few (if any) dependencies, so aiding its composition within applications.

Component frameworks. The use of frameworks helps address the need to publish the interfaces of components, as well as aid loose coupling between components, and the choice of a framework is an important architectural decision.

Software services. Can be viewed as an evolution of components, whereby the services themselves may be provided by third parties and are very loosely coupled to the application, allowing for late binding to particular providers.

Part III

Design as a Verb: Designing Software

Chapter 12

Structuring the Ill-Structured

Having examined the nature of software and some of the characteristics of the design process in the first part of this book, and some of the ways that we can describe these in the second part, in this third part we address the role of '*design as a verb*' and the question of how software might be designed.

We might expect that, in the ideal, the design of software should be an *opportunistic* process, with design decisions being driven by the needs and characteristics of the required application. However, to do so well requires extensive experience and considerable ability from the designer(s). Given that software is often more likely to be designed by less experienced and more imperfect beings, various design strategies and guidelines have been devised to assist them and to help organise the design process.

This chapter introduces these, and then the following chapters (among other things) look at some of them in rather more detail. A core theme that runs throughout is that of how to exchange and share *design knowledge*, whether it be in terms of strategies for developing a design, or ways of organising the design elements. Indeed, given that providing *knowledge transfer* is one of the main goals of this book, this chapter can be considered as introducing some key elements of what this can involve.

12.1 Challenges in creating a design

The difficulties involved in creating software applications and systems have long been recognised by software developers, if not always by others. While other technologies related to computing such as hardware design have raced along, rapidly gaining orders of magnitude in performance, while reducing size and cost, the techniques employed for designing software appear to have inched along in a series of relatively small steps. Much of this of course stems from the ill-structured nature of software design, reinforced by some of the characteristics of software itself. As observed by Brooks (1987) the challenging properties of software include the following.

- *Complexity.* This is seen as being an essential property of software, in which no two parts are alike and an application may possess many states during execution. This complexity is also arbitrary, being dependent upon the designer (and the model they create), rather than the problem.
- *Conformity.* Software, being 'pliable', is expected to conform to the standards imposed by other components, such as hardware, external bodies, or existing software.
- *Changeability.* Software suffers a constant need for evolution, partly because of the apparent ease of making changes (and relatively poor techniques for costing them or recognising the technical debts they create).
- *Invisibility.* As observed in the preceding chapters, any forms of representation used to describe software lacks any *visual* link that can provide easily-absorbed descriptions of the relationships involved. This not only constrains our ability to conceptualise the characteristics of an application, it also hinders communication among those involved in its development (including the customer)

Challenges of software

An interesting question is whether or not these have altered significantly with time. Being intrinsic properties of software itself, both *complexity* and *invisibility* remain issues that are as challenging as they were when Brooks was writing his analysis. Systems remain complex and while design representations have evolved to help cope with different architectures, the lack of any visual parallels remains one of their characteristics. And if anything, the other two properties have come to pose even greater challenges. The connectivity of software systems (typified by the 'internet of things') has increased and software often needs to confirm to standards that themselves change and evolve. Indeed while complexity and invisibility are largely intrinsic properties of software, both *conformity* and *changeability* are more affected by extrinsic factors.

So these properties of software continue to form a set of constraints upon the activities of designers as well as upon the form of the outcome from the design process. And they will all influence the design process, regardless of the scale, purpose and form of an application. In Chapter 4 we looked at some of the knowledge that software designers may possess (or need to possess) as individuals. In this chapter (and the following chapters) we are concerned with the ways in which knowledge about designing software (*knowledge schema*) can be *shared* between experienced designers, and also *transferred* from experienced designers to those having less experience.

12.2 Evolution of knowledge transfer mechanisms

Since the process of software design is an ISP, the very idea of teaching others to design software would appear to be a contradiction. If the design of every software application is both unique and also has many possible forms, we might reasonably ask how we can usefully learn anything other than rather general principles when it comes to acquiring design knowledge. However, while this may be true, it does ignore the fact that, at an abstract level software applications may also have many things in common. Our example of the CCC is focused upon booking the use of some form of resource, in this case cars. But there are many other applications that book resources, whether these be theatre seats, train travel, hotel rooms etc. Individually they will differ, but they will also possess many characteristics in common.

The same effect occurs when we consider architectural style. Here the commonality lies less in what the applications do, and more in the way that their operations are organised (call-and-return, pipe-and-filter etc.). As it happens, the earliest forms of knowledge transfer used with software design were essentially based upon this commonality of architectural form.

Following a plan

Developing a design (implicitly, developing a design model) can proceed in various ways. Early approaches assumed that the designer essentially began with a 'clean sheet' and a clear specification of what was needed. (We may smile at this last assumption, but of course, many early applications such as payroll systems and applications recording commercial transactions *were* relatively well-defined and involved little or no user interaction.) What we usually term *plan-driven* design strategies could be used to create these, essentially by treating the task of designing software as though it were almost a form of WSP, and early forms did so by organising the design process as a sequence of 'steps' that involved:

1. building a model of the requirements (essentially a 'system model') to provide a specification of what was required;
2. mapping this on to the preferred architectural form (typically call-and-return or pipe-and-filter) by allocating the tasks identified in the previous step to specific elements;

3. refining the details of the elements and their interactions (such as the use of parameters) to create the final design model.

We will look at such strategies, their strengths and limitations in Chapter 13. In terms of the properties of software identified above, they do provide support for addressing both *complexity* and *invisibility*, but are less readily able to incorporate ideas about *conformity* or *changeability*.

Early plan-driven approaches commonly employed a top-down strategy with a clear sequence of actions, starting with a description of the system as a whole and then developing a design model that describes the detailed design elements in a step by step manner. As noted above, this may well produce designs that lack flexibility and that can be difficult to adapt and modify. Over the years various forms employing a more 'bottom-up' strategy emerged, culminating in the concept of *agile* development approaches. These aim to provide better support for *changeability* (and to some extent for *conformity*) and are discussed in Chapter 14.

Any form of knowledge transfer tends to depend upon some element of commonality. Plan-driven thinking is largely centred upon commonality of architectural form (essentially implementation). If instead we seek to share ideas based upon commonality of *function*, then this leads to the next development in design knowledge transfer.

During the 1990s, as the use of objects and object-oriented thinking became widely used, there was much greater interest in the idea of *reuse* of design knowledge at different levels. Adelson & Soloway (1985) observed that designers reused past experiences (what they termed 'labels for plans') and as the corpus of software applications grew, the *patterns* concept emerged as a way of codifying useful forms. Whereas a designer was previously expected to begin with a 'clean sheet', basing key elements of a design upon well-established structures or 'patterns' made it possible to draw upon the experiences of others. Again, we look more fully at the ideas involved in Chapter 15, but one thing we can observe here is that one of the strengths of this approach is that it gives greater attention to *changeability*.

The adage that "there is no such thing as a free lunch" does certainly hold true for when designing software. While plan-driven forms may produce design solutions that are relatively constrained and prescriptive, they *do* provide quite a lot of detailed guidance about how to develop them. Later (and 'lighter') approaches may help produce more adaptable and innovative designs, but their use does require greater creativity and skill, as we will see.

12.3 Designing with others

While the preceding sections (and chapters) have tended to emphasise the activities of 'the designer', software is something that we often develop in collaboration with others. Given the nature of the design process, it is hardly surprising that designing software as a team activity adds further complications, some of them in the form of added constraints upon the form of the process, and hence also upon the form of the resulting product. (Of course, one of the hazards of team design is that the number of major components of a system will end up being the same as the number of people in the design team!)

Team design

In (Brooks 1987), Fred Brooks Jr. observed that, at that point in time, many of the very successful and 'exciting' software systems had been the work of one or a few 'great designers'. His examples may now look a little dated (and systems have got much larger, making it less likely for them to be the work of one great designer), but we might still usefully compare the 'clean' design of *Unix* and its derivatives with operating systems such as the *Windows* family.

Given that great designers are rarely available to help us, and that the task of design may have to be shared in some way, we need to consider two issues when designing as a team.

1. How to organise the tasks of the design task among the team, and how to determine how the different elements will interact.

2. How to integrate the individual contributions to the design, which may well require a process of negotiation between the members of the team to resolve different expectations about the overall design.

The first of these can be managed, at least in part, by ensuring that a suitably modular view is taken of the overall design structure. Both will also be influenced by the form of the design team, whether this be a hierarchical one (with someone being the 'chief designer') or a more collaborative one, as typified by Gerald Weinberg's concept of *egoless programming* (Weinberg 1971). The chief programmer model, proposed by Harlan Mills (1971) has the disadvantage of requiring someone with clear authority and knowledge as team leader, so not surprisingly, given the difficulty of finding such people, current ideas such as agile development have tended to put more emphasis upon collaboration and coordination among a set of peers forming a design team.

Researchers have also sought to understand and model the psychological factors that influence team behaviour when designing software (Curtis et al. 1988, Curtis & Walz 1990). Relevant factors they have identified include:

- the *size* of a team—with some indication that a size of 10-12 members is probably an upper limit for productive working;
- the large impact that may be exerted by a small subset of the team who possess superior application domain knowledge;
- the influence of the way that an organisation operates—with particular emphasis upon the need to maintain a bridge between the developers and the customer (Bano & Zowghi 2015).

So, operating as a design team requires the use of knowledge schema drawn from the domain of group psychology as well as technical knowledge about software design. This is perhaps best summarised by the following quotation from Curtis & Walz (1990).

> "*Programming in the large is, in part, a learning, negotiation, and communication process.*"

We will particularly return to this when we examine agile development ideas in Chapter 14.

Experts prefer working with others. PvdH #7

12.4 Empirical knowledge about design creation

Since much of the material of this chapter refers to empirical knowledge about how people and teams create designs, many of the relevant empirical studies have already been discussed.

One topic not really elaborated though is that of *opportunistic* design activities. There are of course good reasons for this. If a process is opportunistic, then its form is determined by the problem, and the way that designers begin to understand both the problem and ways in which it might be addressed. That doesn't offer much in the way of guidelines about how to employ an opportunistic strategy of course, as it doesn't lend itself to being generalised. (Unfortunately, this doesn't therefore fit well with van Aken's idea of *design sciences* that we encountered in Chapter 5, where the goal was to "develop valid and reliable knowledge to be used in designing solutions to problems" (van Aken 2004).)

To highlight this, we briefly examine some outcomes from the study of software design that formed the basis of the book edited by van der Hoek & Petre (2014). This study was based upon recorded observations of a number of pairs of software designers who were addressing a particular problem (creating a simulation package for city traffic management planning). The plot shown in 12.1 is based upon the qualitative analysis reported in Budgen (2014), and was obtained by analysing the recorded utterances of one of the teams of designers and interpreting these to determine which viewpoint was the focus of attention at that point. In all, the utterances of three teams were analysed, and all were very different.

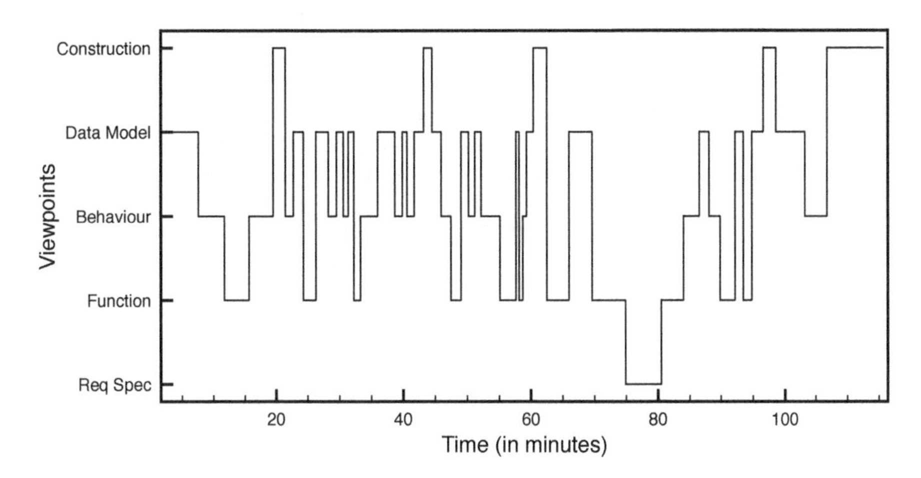

FIGURE 12.1: Example of viewpoint changes during design

The main point is that over a period of nearly two hours, as the two designers in the team produced their design model, their modelling process ranged widely between all four viewpoints, as well as spending some time revisiting the original requirements specification. This was very much an opportunistic design session, with the designers exploring different aspects as their understanding unfolded, and helps to illustrate why such a process cannot readily be generalised.

Key take-home points about structuring 'solutions' to ISPs

Software design practices employ a range of strategies for addressing the ill-structured nature of software applications and need to cope with some important factors.

Challenging properties of software. In 1987 Fred Brooks Jr. identified four key challenges for the designer as arising from the *complexity* of software, the need for *conformity* with other elements of a system, the constant need for *change* and the *invisibility* of software. Complexity and invisibility are largely intrinsic and have remained unchanged, but conformity and changeability are probably much bigger influences (and constraints) than they were at the time that Brooks identified their effects upon the design process.

Knowledge transfer. Mechanisms for sharing knowledge about 'what works' in particular situations have evolved over the years and their evolution is quite closely linked to the evolution of software architectural styles. While plan-driven forms were (and still are) practicable for applications employing call-and-return and similar architectural styles, later forms of architecture require more complex design models and hence more complex processes are involved in developing these.

Design teams. Software is increasingly developed collaboratively within a team, and so the design process needs to provide opportunities for interaction and coordination among team members.

Chapter 13

Plan-Driven Software Design

Plan-driven strategies were one of the earliest mechanisms devised for transferring knowledge about software design. Essentially, the relevant knowledge schema are organised in the form of a structured sequence of well-defined activities, with the activities themselves being performed in a form appropriate to the needs of a particular application.

In this chapter we examine the nature of plan-driven design strategies, and look at some examples of the practices that they employ. As we will see, a significant limitation is that this form of strategy is less readily suited to being used with design solutions that are based upon more complex architectural styles. However, this does not make plan-driven strategies irrelevant. In particular, one of their strengths lies in the way that they 'systematise' the design process. This means that ensuring the integrity of the overall design model during later evolution of an application can be aided by a knowledge of the original method followed. Hence an understanding of their workings can be useful, both as a source of design ideas, and also when seeking to understand and modify legacy systems and applications.

13.1 What does plan-driven mean?

A simple, but often useful, way of transferring knowledge about how to perform a task is to write down a set of general rules that describe a sequence of actions that should be followed. This approach can be particularly effective with non-creative tasks such as assembling an item of flat-pack furniture, but it can also be used to provide guidance for more creative activities.

As a very simple example, the (traditional) way that we make a pot of tea could be performed by following the plan specified below:

> *Boil the kettle; warm the pot; add the tea to the pot; pour on boiling water; wait for a short while; pour the tea into a cup.*

Such a 'method' puts a strong emphasis on the ordering of actions. (If you don't think so, try changing around any of the elements of the above example and see what results!) However, it provides little or no guidance about those quality-related issues that are more a matter of taste, such as:

- how much tea to add
- how long to wait for the tea to brew

since these are decisions based upon the personal preferences of the person making (or consuming) the tea as well as being dependent upon the type of tea.

Because they can be considered as being equally essential to the success of the tea-making process, we can reasonably assume that 'sequence' knowledge' alone is insufficient for even such a relatively simple task, and that some form of 'domain knowledge' is also required.

Making tea

Cookery books provide a particularly good illustration of this. A recipe may provide details of certain tasks that should be done, together with a schedule (or ordering) for doing them. When followed by a novice it can be hoped that this will result in some degree of success, while the more accomplished and creative reader can interpret and adjust it to create more impressive effects. (Well, that's the basic thinking anyway.)

We can find many other examples in everyday life—the instructions followed by a hobbyist to create a model steam engine; dress-making patterns, a guide to planning a walk in the mountains etc. Taken together, these suggest that procedural guidance can also be an effective way of guiding others to perform creative tasks. However, in the first two examples the scope of the creative element is constrained—the recipe describes one particular dish, the dress-making pattern produces a particular garment. In contrast though, procedural guidance used to advise about planning a walk in the mountains will necessarily be less prescriptive, since it needs to be adapted for routes of different lengths and difficulty, as well as factors such as those arising from the size of a group etc. Hence the knowledge schema provided by such a form will require the user to employ additional expertise and complementary forms of knowledge when doing the planning (which as we observed earlier, is a form of design activity).

Plan-driven software design strategies have essentially adopted this model in order to provide guidance on designing software applications. The resulting design strategies have often been described as 'methods', implying a structured and procedural way of doing things.

However, we do need to appreciate that when a plan-driven approach is adopted for use with more creative tasks, such as planning walks or designing software, it is effectively operating at a higher level of abstraction than when used for describing how flat-packed furniture should be assembled. It is rather like employing a *meta-recipe* approach to provide guidance on how to produce a wide range of recipes, which requires much more creative input in order to adapt it to different dishes. Hence, while a plan-driven approach can provide good guidance for a task that involves a highly constrained degree of creativity, such as making tea (or coffee), it will necessarily be much less prescriptive when individual design decisions need to be driven by the needs of the problem.

Understandably perhaps, plan-driven software development is now often viewed as being mainly of historical interest. On the debit side, it does have two significant limitations: one is that it is implicitly tied to an overall 'waterfall development' context; the other is that it is more likely to be effective with less complex software architectures. On the positive side, many applications have been successfully developed using a plan-driven strategy; and some important concepts (such as *coupling*) have emerged from its use. And of course, as mentioned earlier, an understanding of how such strategies are organised can help with preserving the integrity of the resulting design during maintenance.

13.2 Decompositional and compositional strategies

While an *opportunistic* strategy may commonly be used by experienced software designers (Visser & Hoc 1990), such an approach cannot readily be incorporated into a plan-driven structure. Plan-driven methods therefore organise their guidance around one of two major problem-solving strategies.

13.2.1 Top-down decomposition

Decomposition

When early software applications were being developed, a common choice of architectural style was one of call-and-return, since most implementation languages had features that mapped closely to the structure of the underlying machine, aiding generation of efficient code. Applications therefore employed some form of 'main' program unit (often incorporating any permanent and global variables) and a set of sub-programs, with these performing their tasks using local variables that were only in scope while the sub-program was executing. So early thinking about design was focused upon how to map application functionality on to such a structure.

In a pioneering paper, Niklaus Wirth (1971) proposed the use of a process in which the task to be performed by the overall problem was gradually *decomposed* into a set of smaller ones (termed *stepwise refinement*). As envisaged by Wirth, both functionality and also data were to be refined in this way, and duly mapped on to the main program and sub-programs. Among the lessons that he derived from this model were two that we might usefully note here.

- The modular structure resulting from this process determines the ease with which a program can later be adapted to meet changes in the requirements or its context. Although this idea is demonstrated in his paper, the changes illustrated in his example were ones that largely extended the functionality involved in the original problem posed, and we should note that the ideas about *information hiding* later expounded in Parnas (1972) provide a much more coherent strategy for determining the choice of modular structure.

- Each refinement in the process of decomposition embodies a set of design decisions that are based upon specific criteria. This reflects the recognition that, as we have already observed many times in this book, designing software is not an analytical process, with a need to assess (and re-assess) the influences of a range of factors at each design step.

Wirth's paper was undoubtedly a seminal one, but although his ideas were centred upon the concept of *modularity*, his paper provided no criteria that a designer could employ to compare between possible choices of module structure (Shapiro 1997). However, the concept of module *coupling* introduced in

(Myers 1973), that later formed one of the underpinnings of the 'structured design' school of thinking (Stevens et al. 1974) that are examined later in this chapter, did provide criteria that could be used to assist the process of stepwise refinement, and so coupling became an underpinning concept used as part of this approach.

Myers argued that the "primary goal in designing a modular program should be to decompose the program in such a way that the modules are highly independent from one another". He advocated the use of 'strong modules', which performed a single, well-defined function and he identified some of the ways in which coupling between modules could arise.

13.2.2 Compositional design strategies

A rather different, and often more challenging, way of creating a design is to use a *compositional* strategy. Whereas the decompositional approach tends to focus upon creating the structure of the resulting application around consideration of its *function*, usually realised through by the operations that an application needs to perform, a compositional approach seeks to create a design model that is formed from the descriptions of a set of distinctive *entities* or components that can be recognised in the problem, together with descriptions of the *relationships* that link these entities. The nature of the entities and the relationships will vary with the method and the architectural style that is adopted. Indeed, whereas decompositional forms are effectively constrained to producing design models using an architectural style such as call-and-return, compositional strategies can be used to create design models for a variety of architectural styles, including those using processes and objects.

Employing a compositional approach is usually considered as being less intuitive than the use of top-down decomposition (Vessey & Conger 1994). In part this may be because of its greater complexity, requiring a more mixed set of viewpoints for creating the design model, and also because it provides more scope for opportunistic decision making. However, it can be argued that the use of this strategy is likely to lead to more consistent design solutions, regardless of who is doing the design, since the design strategy aims to relate the structure of the 'solution' (design model) to that of the 'problem' rather than the structure of the underlying machine (Détienne 2002). The process of compositional design also places an emphasis upon 'grouping' elements when elaborating the design model, where groupings can be based on criteria such as the different forms of coupling, which again may be related closely to both the problem as specified in the requirements and the chosen architectural style.

Composition

13.3 What do plan-driven methods provide?

One question we should ask at this point is why should anyone want to use a plan-driven method? Probably far more software has been developed without using anything in the nature of a 'method' than through the use of such methods, so what benefits does their use confer? To answer this, we first need to look at the mechanisms they employ in a little more detail.

Vessey & Conger (1994) suggested that the knowledge involved in following a plan-driven strategy can be categorised into two forms:

1. *declarative knowledge*, which describes the tasks that should be performed at each step in the process; and

2. *procedural knowledge*, consisting of knowledge about how to employ a given 'method' in a particular situation.

In terms of our example of making tea, the declarative knowledge involved would consist of an understanding of the tasks to be performed and of the order in which they should be performed, while procedural knowledge would address such issues as how much tea to put in the pot and how much water to add in order to make tea for four people.

Declarative knowledge schemas can be codified by providing a 'do this, then do that, then follow this by doing...' form of description, whereas procedural knowledge schemas are more likely to be acquired from experience and best codified in the form of advice. Since we often express the declarative knowledge in what we might term a procedural form, by specifying the sequence of actions that should be performed, this terminology can also be a little confusing. (And it is worth noting that in (Détienne 2002) the terms declarative and procedural are used in yet another, slightly different, way.)

As used for software design, plan-driven methods generally embody the relevant design knowledge through the use of three main elements.

- The *representation* part consists of a set of notations that can be used to describe (or model) both the characteristics of the original problem and also that of the intended 'solution' (the design model), using one or more viewpoints and different levels of abstraction.

- The *process* part provides the declarative knowledge by describing the procedures to be followed in developing the design model, and suggesting strategies that can be used and adopted when making any design choices involved. Following this usually involves the designer in making one or more transformations between the different notations that comprise the representation part.

- A set of *heuristics* or *clichés* provides guidelines on the ways in which the activities defined in the process part can be organised or adapted for particular classes of problem. These may well be based upon experience of past use.

In terms of the classification employed by Vessey & Conger (1994), the process part can be considered as embodying declarative knowledge, while heuristics provide a means of supplying procedural knowledge. (The representation part is a vehicle for capturing knowledge about the design model rather than a mechanism for creating it.) We will use these three elements as a framework for discussing how the examples of plan-driven forms discussed in the rest of this chapter are structured.

However, to return to the question posed at the start of this section, namely what benefits do users hope to obtain from using a 'systematic design method', it can be argued that some of the benefits are as follows.

- The representation part of a design method provides an artificial framework to help with thinking about an invisible set of elements. Portraying ideas about something that is invisible is inevitably a challenging issue when designing software, and the representation part of a method does provide a syntax and semantics to aid this through the notations adopted.

- Design methods can assist with managing some of the effects of *scale* and of the cognitive load this imposes, particularly where teams are involved. They do so by ensuring that consistent forms of description are used for sharing knowledge among a team. Recording design plans in a consistent manner may well be important for future evolution (and while they may not necessarily reduce the *technical debt* involved, they can at least help with making it explicit). However, while design records may help a maintenance team to understand the intentions of the original developers (Littman et al. 1987), as Parnas & Clements (1986) have observed, such documentation may well be more helpful if it describes an 'idealised' process rather than the more opportunistic strategy that may well have actually been followed, even for a plan-driven method.

- As already noted, plan-driven methods act as a knowledge transfer mechanism. Observational studies suggest that although experienced designers may often work in an opportunistic manner, this practice may be less well-formed and reliable when the designer is less familiar with a problem or its domain (Adelson & Soloway 1985, Guindon & Curtis 1988, Visser & Hoc 1990). So for the inexperienced designer, or the designer who is working in an unfamiliar domain, the use of a systematic design method may help with formulating and exploring a design model. Hence to some extent at least, *method knowledge* may provide a substitute for *domain knowledge*, where the latter is inadequate or lacking.

Figure 13.1 is a revised form of the 'problem-solving' model that we originally saw in Figure 1.1. Part (b) has been revised to show how a plan-based design strategy addresses the problem of 'solving' an ISP (and is much simpler than the form described in Figure 1.1). We can also see that in some ways, the problem-solving process in Figure 13.1(b) is much closer to the form in Figure 13.1(a), which describes how we solve WSPs. Of course, this is rather idealised, since iteration may well occur between the steps of a method, but nonetheless, the design process involved is a much more constrained one.

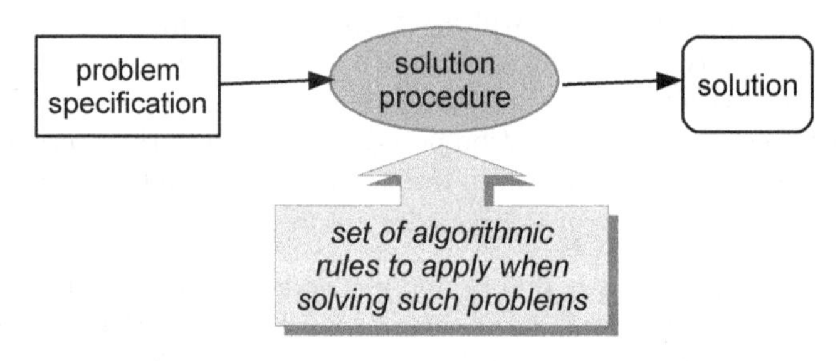

a) Process for 'solving' a WSP

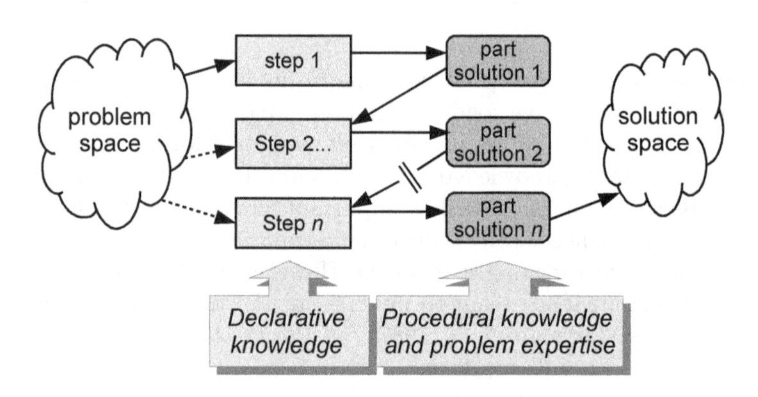

b) Plan-driven process for 'solving' an ISP

FIGURE 13.1: Addressing an ISP using a plan-driven form

However, while a systematic design method may help to structure the strategy used to address an ISP, it cannot magically turn it into a WSP, and in particular, the designer's choices and decision-making will still need to be made on the basis of the problem being addressed. What a design method assists with, is guiding the designer about determining when it might be appropriate to make those decisions.

It may be helpful here to return briefly to the analogy of the recipe for cooking. A recipe describes the process that should be followed in order to create a particular dish, in much the same way as a software application may embody the algorithm needed to rank the positions of the nearest cars for the CCC. Designing software applications is therefore rather like designing recipes, it requires the designer to possess insight into the problem domain; to be aware of the potential that is present in the various available materials; and to appreciate the bounds upon the software application (or cook) that will perform the task. So plan-driven (and other) design methods are essentially *strategic* in their scope rather than prescriptive.

> *Experts use design methods (selectively). PvdH #17*

The rest of this chapter looks at some examples of plan-driven forms. Of necessity the descriptions have to be brief (a full description for any of them merits a book in itself). However, they should be sufficient to show how plan-driven strategies have evolved and something of the way that they can help with addressing the problem of designing software.

13.4 SSA/SD: example of an early plan-driven form

The *structured systems analysis/structured design* (SSA/SD) method in its various flavours provides a good example of an early plan-driven approach to design. It is one that evolved over a period when software applications were getting larger and ideas about how to design software were in a rapid state of evolution. SSA/SD was by no means the only approach to software design being explored and documented in this period, but became quite a widely-used form, perhaps because it stemmed largely from work performed by IBM, which was then a leading organisation involved both in developing software and also researching into the associated issues. Various authors published books describing forms of SSA/SD (Gane & Sarsen 1979, Page-Jones 1988, Connor 1985), as well as variants such as real-time applications, covered in the books by Hatley & Pirbhai (1988) and Ward & Mellor (1985). For this section we will largely follow the form described by Page-Jones (1988).

As a design method, this one is really a composite of two separate but related techniques. *Structured systems analysis* is concerned with modelling the problem-related characteristics of the application, making use of a set of representation forms that can also be used for architectural design. *Structured design* is then concerned with the 'solution'-related aspects involved in producing a design model.

In its early forms, the basic design strategy was a refinement of top-down design, with the choices involved in the functional decomposition process being moderated and constrained by considerations of information flow, and to a lesser degree, of data structure. Subsequent variants adopted a more compositional approach for the analysis stages, based upon such techniques as *event partitioning* (Avison & Fitzgerald 1995). There were some later developments that tried to adapt the approach to an object-oriented style, but these do not seem to have been adopted very widely, and apart from these, the evolution of SSA/SD forms effectively appears to have come to an end in the later 1980s at a time when ideas about software architecture were evolving rapidly.

In the rest of this section we briefly examine the representation part, process part and some of the heuristics used in this method.

13.4.1 SSA/SD representation part

The two techniques make use of quite different notations. The *structured systems analysis* element primarily builds a problem model using data-flow diagrams or DFDs (see Section 9.2), while the *structured design* component creates a design model around the use of structure charts (see Section 9.5).

The functional model created by using the DFDs can be augmented through the use of more detailed descriptions of the bubbles in the form of 'process specifications' or *P-Specs*, where these can be regarded as providing a subsidiary functional viewpoint. A P-Spec is a textual description of the primitive process represented by a bubble, summarising the process in terms of its title, input/output data flows and the procedural tasks that it performs.

An additional element of the 'problem model' is a *Data Dictionary.* This can be used to record the information content of data flows, drawing together descriptions of all of the data forms that are included in the DFDs, P-Specs and any other elements that might be used. In its initial form this should be highly abstract and avoid focusing upon physical format (Page-Jones 1988).

Whereas problem modelling can use a variety of supplementary forms (such as ERDs and state transition diagrams), the solution modelling part is confined to using structure charts. As observed earlier, the structure chart is a program-oriented form of description that maps on to the call-and-return style.

13.4.2 SSA/SD process part

The process embodied in this method can be regarded as an elaboration of the basic top-down approach based on functional decomposition, being extended by adding such elements as consideration of information flow. A simple description of the process is as follows.

1. Construct an initial DFD (the *context diagram*) that provides a top-level description of the application in terms of a single bubble together with a set of inputs and outputs.

2. Elaborate the context diagram into a hierarchy of DFDs, and while doing this, develop the accompanying data dictionary.

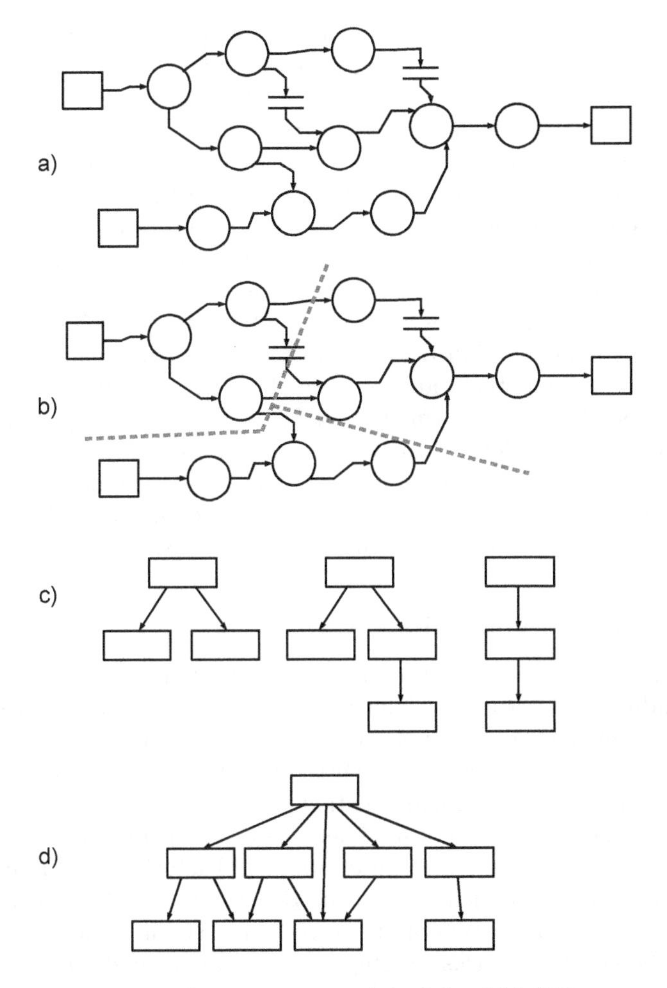

FIGURE 13.2: A schematic model of the SSA/SD process

3. Use *transaction analysis* to partition the DFD into tractable units that each describes a specific 'transaction'.

4. Perform a *transform analysis* upon each of the DFDs resulting from Step 3, in order to produce a structure chart for the corresponding transaction.

5. Merge the resulting structure charts to create the basic design model, refining them as necessary to address issues such as initialisation, error-handling and other exceptions that are not part of the main operational activities.

Figure 13.2 provides a schematic model of this process. In part a), we see the complete system DFD produced from steps 1 and 2. Part b) shows the process of transaction analysis, with the DFD being partitioned into three smaller DFDs that address specific tasks (the dashed lines show the divisions). In part c), transform analysis has been used to create structure charts for the three transactions, and then finally in part d), these have been combined to create the complete design model for the application program. (This is very idealised, and we have assumed here that the functionality in each bubble can be readily mapped on to a single sub-program.)

Each of these steps involves what we might consider to be design activities (use of the term 'analysis' in this method can be a bit misleading). Steps 2-4 form the core of the method, and as we might expect, while there is guidance available about how to perform these tasks, actually mapping them on to a real problem is a task 'left for the reader'. Transform analysis in particular is where the design model becomes modified, and while the emphasis upon data and function is retained, the logical model embodied in the DFDs is changed into a 'physical' model that maps on to sub-programs.

13.4.3 SSA/SD heuristics

Heuristics are often created when a particular design method is used on a set of broadly similar problems. This does not seem to have been the case with SSA/SD although a number of heuristics have become established to provide guidance on such tasks as creating a DFD, or performing a transform analysis.

A technique that we should mention here is that of *factoring*, used to separate out the functions of a module, where the operations of one module are contained within the structure of another. Factoring can be considered as a reflective activity aimed at reducing the size of modules, clarifying their features, avoiding duplication of operations and helping with reuse of design elements. Some of the tasks of Step 5 above are related to factoring and together they highlight the way that the use of such design methods may still depend upon an additional design element in order to produce 'efficient' design models.

13.5 SSADM: a designed design method

Our second example is a plan-driven design method that has rather different origins, and hence provides some interesting features. SSADM (structured systems analysis and design method) was developed by Learmonth & Burchett Management Systems on behalf of the UK government. It was intended to support the development and long-term maintenance of data-intensive systems since the use of these, and hence their systematic production, was seen as playing an important role for both central and local government agencies.

Part of the rationale for commissioning such a development was the way that public bodies in the UK are traditionally organised. In particular, central and local government agencies often move staff to new posts on a regular basis (the UK civil service has tended to favour the idea of the 'generalist' over the 'specialist', and hence staff development tends to require the gaining of wide experience). Where such roles involved the responsibility for software development, it was considered that regular change of staff could (and did) lead to discontinuity in the way that the software applications were developed, and hence that using a mandated approach for their development would help overcome this.

SSADM Version 3 was the first stable version of the method, and was supported by a small number of textbooks. This evolved into Version 4 in the early 1990s, at which point the method was considered to be 'mature', and which appears to have marked the end of development for both SSADM and for the textbooks supporting it (Longworth 1992, Weaver, Lambrou & Walkley 2002). This period also coincided with the growing uptake of object-oriented platforms, which would have been likely to compete as an alternative strategy for use in government projects.

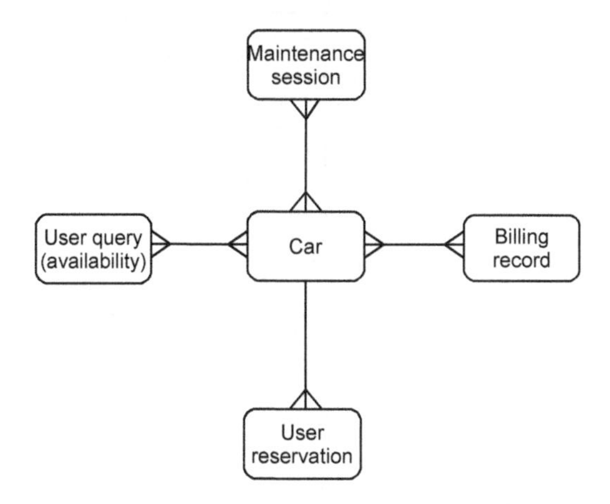

FIGURE 13.3: Example of an LDS

13.5.1 SSADM representation part

The forms adopted for SSADM largely reused established ideas and notations, albeit with some (sometimes rather confusing) renaming and relabelling. SSADM analysis and modelling practices address the following main viewpoints.

- *Data relationships.* These are modelled using the fairly conventional entity-relationship diagram, relabelled as a *logical data structure* (LDS), with the modelling process being termed 'logical data modelling', or LDM.

- *Data flow.* This is modelled using a form of DFD, which is either developed by transformation of an existing application or through decomposition. However, these are more solution-oriented than the 'bubble' forms used by De Marco and others for analysis.

- *Function and behaviour.* Modelling of these viewpoints uses *entity-life-history* (ELH) diagrams that have the same form and syntax as a Jackson structure diagram (Jackson 1975).

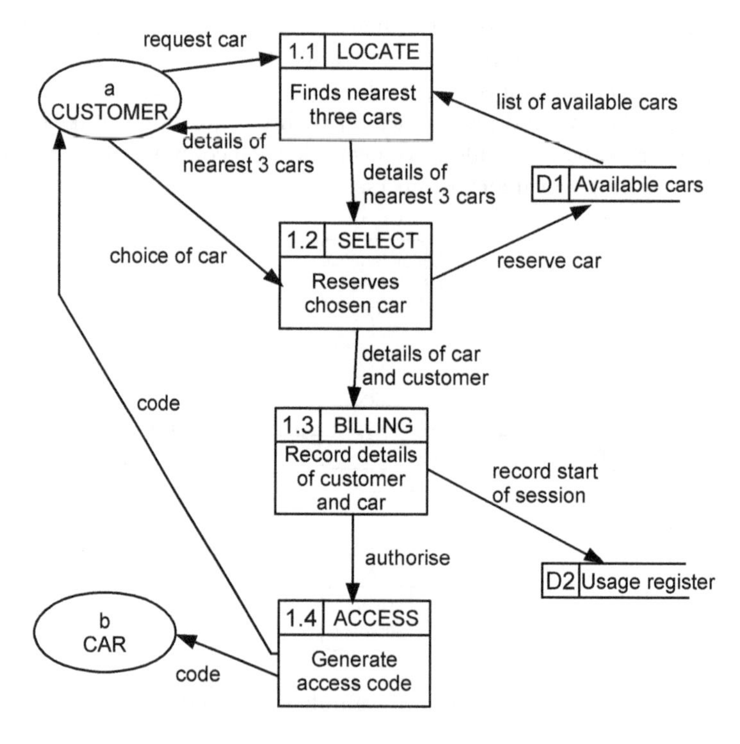

FIGURE 13.4: Example of a DFD using SSADM notation

Figures 13.3 to 13.5 provide examples of each of these forms. Within the Jackson structure diagram notation, a 'plain' box signifies a single action within a left-to-right *sequence*, a box with a star in a corner indicates *iteration* of the action (including the possibility of no action), and boxes with circles indicate *selection* between optional choices. For the latter, an unlabelled box indicates that one option might be 'do nothing'. At any level of expansion of

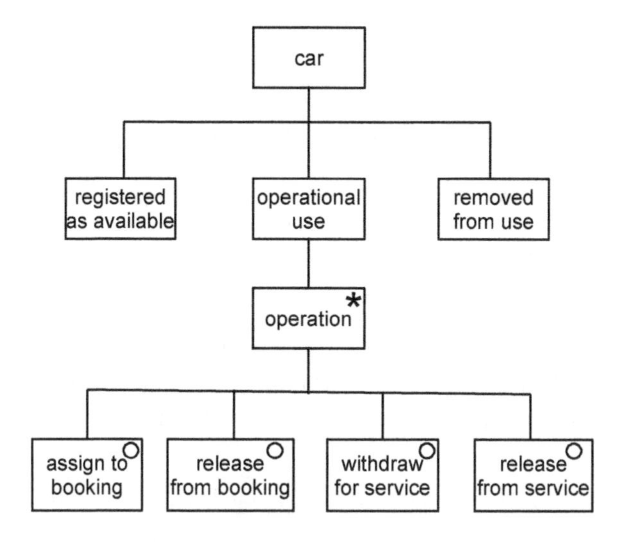

FIGURE 13.5: Example of an entity-life-history diagram

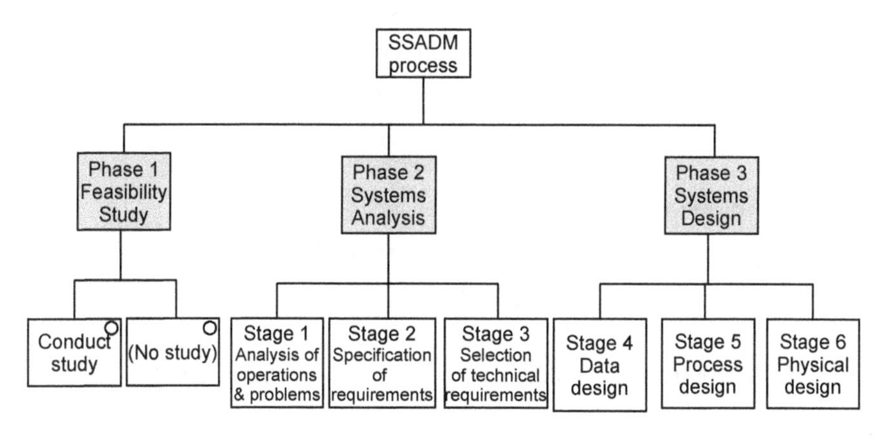

FIGURE 13.6: The SSADM process part

a branch of the model, only one form (sequence, iteration, selection) can be used—they cannot be mixed.

13.5.2 SSADM process part

Declarative knowledge is embedded into SSADM through a strongly hierarchical process, with three top level *phases*. The first (a *feasibility study*) is optional, and the other two phases each consist of three *stages*. The stages are further sub-divided into *steps* which in turn each encompass a set of one or more *tasks*. Figure 13.6 uses the Jackson structure diagram format to describe the top levels of the process part.

SSADM practice is very much focused upon the idea of producing a *logical design* for an application, and this is only mapped on to a *physical design* in the final stage. So, in principle at least, there is no strong coupling to a particular architectural style. SSADM is also a very bureaucratic method, as might be expected from its origins and purpose. Figure 13.7 provides an example of Step 320 (the first digit indicates which phase of the method this concerns). Note in particular that there is also a strong emphasis upon *documentation*, which again has an important role in this method. However, when we examine empirical knowledge about plan-driven forms in Section 13.7 we will see that the procedures of SSADM can be, and have been, used in part without necessarily employing the complete method.

Inputs:	Option Specifications from (step 310)
	Required system specification (from stage 2)
Tasks:	1. Prepare presentation plan
	2. Prepare presentations for each option
	3. Deliver presentations
	4. Provide assistance and advice for option selection
	5. Note decisions of users
Techniques:	User option selection
Outputs:	Note of option decisions (to 330)
	Option specifications (to 330)

FIGURE 13.7: Step 320: user selects from technical options

13.5.3 SSADM heuristics

SSADM can be regarded as using rather formalised heuristics in a number of the steps, referring to these as *techniques*. There are 13 of these, and their main role is to provide procedural knowledge that can be used to help develop the diagrams, with SSADM placing heavy emphasis on the use of *matrices* for this purpose. A matrix in this context is a grid or table that helps with identifying links between elements, rather as the state transition table described in Section 9.3 can be used to develop state transition diagrams.

However, these could also be regarded as providing declarative knowledge as well as procedural knowledge, although it can be argued that their use augments the SSADM process rather than forming a part of it. Figure 13.8

shows an *entity-life-history matrix* (ELH matrix) that can be used to help produce the ELH diagram shown in Figure 13.5.

Entities	Events				
	seek nearby cars	reserve car	start session	release car	service withdrawal	service release
car	*	*		*	*	*
customer	*	*	*	*		
booking system	*	*			*	*
billing system			*	*		
maintainers					*	*

FIGURE 13.8: Example ELH matrix

13.6 Plan-driven design for object-oriented models

Early thinking about how to design object-oriented systems not unnaturally sought to use experience from the approaches used to design applications that used call-and-return and other forms of software architecture. Since plan-driven forms were then the main established vehicle for knowledge transfer, it was not surprising that design 'methodologists' sought to extend the use of what was generally considered a generally effective approach. Indeed, many 'first-generation' OO design methods made varying degrees of use of existing ideas and notations.

However, while in principle, plan-driven approaches are a viable means of providing knowledge transfer about OO design, the greater complexity of the OO model, when compared to architectural styles such as call-and-return, has tended to result in a correspondingly greater complexity of structure for associated plan-driven approaches.

This section examines two OO plan-driven methods (*Fusion* and the *Unified Process*). Both represent attempts to create a second-generation plan-driven method by merging ideas from first-generation OO methods that were seen as effective. In many ways, the UP can also be considered as the point at which the use of a plan-driven form both achieved maximum complexity, and also started to morph into more agile forms.

13.6.1 The Fusion method

As indicated above, *Fusion* can be viewed as being a 'second-generation' OO method. Indeed the methodologists at the Hewlett-Packard Laboratories in Bristol UK who developed Fusion described their goal as being to "integrate the best aspects of several methods" (Coleman, Arnold, Bodoff, Dollin, Gilchrist, Hayes & Jeremes 1994). Among the problems intrinsic to an object-oriented approach that they identified as drivers for their work, the method developers included the following.

Difficulty with finding the objects. They observed that "finding the right objects and classes in an object-oriented system is not easy".

Function-oriented methods being inappropriate. Their observation here was that for OO applications, the traditional methods of analysis and design (that is, those based on data-flow and function) "no longer work". Implicitly, the use of these was seen as a weakness of many first-generation approaches. In particular, they noted that "functional decompositions clash with the object structure of the implementation".

While Fusion seems to have had some early success in terms of its adoption, it never seemed to acquire a solid base of users. The text by Coleman et al. (1994) remains the only readily available description of the methods, and seems to be mainly cited when authors are discussing OO methods. Wieringa (1998) did note that there had been some subsequent development of the method, but that these were not readily accessible, and so the description provided here is that described in the 1994 version. Since the more distinctive element of Fusion lies in the process part, we describe this first (there are no evident heuristics).

The Fusion process

Reflecting the need for more complex processes to address the more complex architectural style created by the use of objects, the Fusion process has no fewer than eight distinct steps. Four of these are classified as *analysis* (black box modelling) while the other four are white box design steps. Continuity through the overall process is provided through the use of a *data dictionary*, which is intended to act as "a central repository of definitions of terms and concepts". Arguably, this is the one element that is most evidently (and sensibly) carried over from earlier structured design practices. The eight steps are shown in Figure 13.9, and can then be summarised as follows.

1. **Develop the object model.** While the task of this step is identified as identifying a set of candidate classes, relationships and attributes, the method does not specify how this should be done. The outcomes are a set of entries in the data dictionary, plus a set of diagrams making up an initial object model for the system as a whole, together with its environment. Advice on conducting this step includes using brainstorming and *noun-verb* analysis[1].

[1] Noun-verb analysis was used in some first-generation OO methods. It involves producing a written description of a 'rough' design solution and then identifying candidate objects from nouns and candidate operators from the associated verbs.

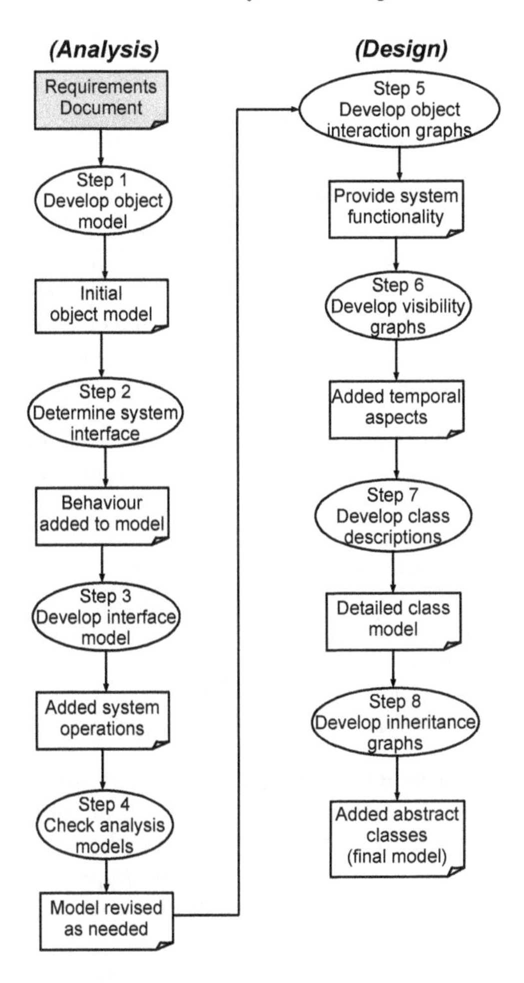

FIGURE 13.9: The *Fusion* design process

In addition to identifying candidate classes, relationships and attributes, this process should also identify any *invariants* and record their details in the data dictionary. For our purposes an invariant is some form of constraint upon the overall state of the system.

2. **Determine the system interface.** The model from step 1 is essentially a static one, and this step extends it to describe the dynamic behaviour of the system as a whole. In doing so, it also clarifies the system bounds by determining what is, and what is not, included in the system itself.

   The core concept here is that of the *event*, with the designer being encouraged to think of how the system is to respond to external events as well as to consider what events will be generated in response to these.

The approach suggested is to consider a set of *scenarios* of use, modelled as *timeline diagrams* (essentially message sequence diagrams, as described in Section 10.6), extending the data dictionary to include these scenarios.

3. **Development of the interface model.** The first element of this involves creating a set of use cases that generalise the scenarios from the previous step (termed *life-cycle expressions*). These are then used as the basis for the 'operational model' that defines the semantics (meaning) for each system operation, expressed in terms of informal pre-condition and post-condition specifications. Overall, this provides a functional description of the system.

4. **Check the analysis models.** This step is intended to ensure *completeness* and *consistency* in the (still black-box) model, rather than extending it in any way. Completeness is largely a matter of cross-referencing with the requirements specification (a limitation of plan-driven models is that they implicitly require a requirements specification as their starting point). Consistency is concerned with ensuring that the different viewpoint models from the previous steps (which we can categorise as being constructional, behavioural and functional) as well as the invariants represent the same overall design model. This can be partly achieved by using scenarios to trace event paths through objects.

5. **Develop object interaction graphs.** The first step categorised as being 'design' is used to describe the designer's intentions about how the objects will interact at run-time to provide the required functionality. This involves creating an *object interaction graph* for each system operation, and in doing so, making decisions about message passing between objects and any resulting state changes.

   One of the elements also clarified in this step is the point at which objects should be created from the classes. (Although the descriptions tend to be couched in terms of objects, since these are the executable elements of the eventual system model, the model itself is really described in terms of classes—while recognising that many classes will only lead to the creation of a single object.)

   A question here is whether the set of objects used in this step is essentially that created in Step 1? The answer to this is 'yes', as the objects involved are defined as design objects that are derived from the analysis objects. So these are still relatively high-level elements that may later need to be expanded or refined.

   Each object interaction graph has a single *controller object* which has a message entering from outside of the particular set of objects involved in the operation, and also one or more *collaborator objects* that make up the rest of the set. There are two important design decisions here:

one is identifying the controller, while the second is the possible need to introduce new objects (classes) to provide the required functionality.

6. **Develop visibility graphs.** These are used to describe the other classes that a class needs to reference, as well as the forms of reference to be used, and their lifetimes (whether the knowledge should persist through the life of the system, or be more transitory).

   This temporal aspect is an important (and distinctive) feature of this Fusion step. However, at this point, the influence of time is largely confined to its influence upon the constructional aspects.

7. **Develop class descriptions.** This step involves drawing together the original object model with the outcomes from the previous two steps to produce class descriptions that specify the following characteristics for each object.

   - The methods and their parameters (derived from the object interaction graphs and object visibility graphs).
   - The data attributes (from the object model and the data dictionary).
   - The object attributes (largely from the object visibility graph for the relevant class).
   - The inheritance dependencies.

   There are clearly quite a lot of decisions related to different forms of *uses* relationships in this step, and it is really the first time that inheritance structures are significantly involved (although there is provision for identifying these when creating the original object model in Step 1).

8. **Develop inheritance graphs.** While the previous step was concerned with ideas about specialisation through inheritance derived from the original object model, this step is concerned with identifying new abstract superclasses. In particular, the designer is encouraged to look at the relationships between the classes developed in Step 7 and identify common abstractions.

While some elements of Fusion are relatively loosely structured, such as the model building in Step 1, which can be expected to be strongly problem-driven, the subsequent steps provide quite good guidance on refining this. Fusion does make good use of diagrammatical notations to develop the functional and behavioural viewpoints and also to ensure consistency between the different viewpoints.

The Fusion notations

Fusion makes quite extensive use of box-and-line notations, although with a much less complex variety of forms than is used in the UML. The syntax is also less detailed than that of the UML.

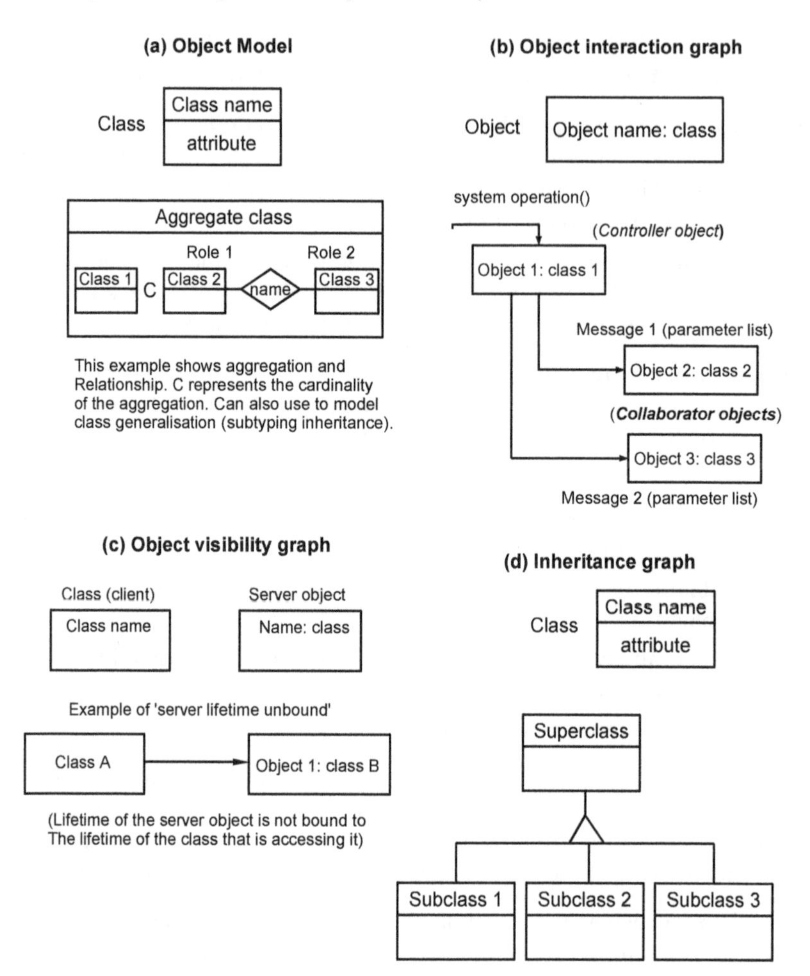

FIGURE 13.10: Some Fusion notation forms

Like other OO methods and the UML, the construction viewpoint employs a class diagram (termed the *object model* that has strong similarities to the ERD notation described in Section 9.4. A simple example of this is shown in Figure 13.10(a).

Class *behaviour* makes use of textual descriptions. Informally, in (Coleman et al. 1994) there are examples that describe scenarios by using a form similar to the UML *message sequence diagram* that we described in Section 10.6.

In the later, more detailed stages of model development, *functionality* is described by using *object interaction graphs* to describe the way that objects collaborate. A graph is created for each system operation. Figure 13.10(b) shows an example of this form, which does differ rather from other notations encountered so far.

The second form of notation that can perhaps be considered as being specific to Fusion is the *object visibility graph*, shown in Figure 13.10(c). This is used to describe the 'reference structure of classes in the system', identifying the objects that class instances need to access and the forms that these references should take (including their permanency or otherwise). In effect, this notation is intended to describe encapsulation, and hence can be regarded as providing a data-modelling viewpoint.

Finally, Fusion makes use of a relatively conventional *inheritance graph*, which again, provides support for modelling of the constructional viewpoint. An example of this is provided in Figure 13.10(d).

Fusion—some observations

The description above, brief as it is, should be sufficient to allow us to make a number of observations about the use of plan-driven approaches to create an object-oriented model.

1. Fusion employs a quite complex four-viewpoint model at a relatively early stage in the design process, particularly when compared to the simpler models employed by SSA/SD in the early stages.

2. The process throughout is largely one of refinement and elaboration, possibly because of the more complex design model, rather than involving any element of transformation between viewpoints (a key element of SSA/SD).

3. The basic set of candidate objects (a major design decision) is largely determined at the very start of the process. A benefit of this is that the design options are thereby constrained from an early point, while a disadvantage is that it is then necessary to get the 'right' object model established very early in the process. For less experienced designers, this can be a significant challenge.

4. The characteristics described by the constructional viewpoint play a much more 'up front' role than in more traditional plan-driven approaches.

5. While the concept of inheritance is integrated into Fusion, it is restricted to appearing very early (seeking to recognise domain-based opportunities) or much later, looking for constructionally-based opportunities.

In terms of the characteristics of the object model, it can be argued that Fusion does provide support for all of the major ones: abstraction, modularity, encapsulation, and hierarchy. Fusion also handles the often quite difficult distinction between the class and the object quite effectively. When considering static and abstract issues, and specifying general characteristics, then the emphasis is rightly upon the *class*. When considering system behaviour, as well as temporal and dynamic characteristics, then the *object* is probably the better abstraction to employ. As a method, it does keep these distinct, and encourages the designer to use whichever is the more appropriate in the specific steps.

Fusion therefore demonstrates that the use of a fairly 'traditional' form of plan-driven approach appears to be feasible when designing object-oriented systems. However, key decisions about the choice of objects need to be made at a very early stage. Indeed, this need to identify key elements early is probably a disadvantage of all plan-driven approaches, regardless of the architectural style employed.

13.6.2 The Unified Process (UP)

The *Unified Process* (UP) stems from the work of the 'three amigos': Grady Booch, James Rumbaugh and Ivar Jacobson. It draws strongly upon early work by Jacobson at Ericsson, and his later development of the *Objectory* (*Object* Fact*ory*) method. The UP also exists in more commercial forms, with the best known of these being the *Rational Unified Process* (RUP).

The authoritative text on the UP is Jacobson, Booch & Rumbaugh (1999). Two other widely-cited sources are by Kruchten (2004) and Arlow & Neustadt (2005). Perhaps because the UP is closely associated with the UML, there are also various texts on combinations with other forms, such as design patterns, described in Larman (2004).

While, like Fusion, the UP represents a merging of ideas from many sources, it differs from Fusion in two significant ways.

1. The sources for the UP have included some of the most popular of the available object-oriented methods and notations, as observed by Johnson & Hardgrave (1999).

2. The resulting process structure is much less sequential than that employed in Fusion. Its form comes much closer to RAD (Rapid Application Development) methods such as DSDM (described in Chapter 14) that represent something of an interim form between plan-driven and agile approaches.

A consequence of the first difference is that the association with the UML has meant that it uses forms that are, at least in part, more familiar to users. The consequence of the second is that it can probably be regarded as something of a 'bridge' between plan-driven and agile ideas. While still more structured than agile forms, it is nonetheless more iterative in terms of the processes involved than those employed by plan-driven forms. Taken together, these may at least partly explain why the UP has continued to be of interest into the 2000s, while Fusion has largely receded into the shadows. (This is not a comment or assessment about relative technical excellence, more an observation that the form of the UP has probably fitted better with the way that ideas about software development have evolved.)

The process of the UP is organised in a far less linear form than conventional plan-based methods, although of course there is an overall flow from beginning to end of the process. The main elements are as follows.

- Four project-based *phases* of development (inception, elaboration, construction, transition) that each completes a major milestone in a project.
- Each phase is organised as a set of one or more *iterations* or 'mini-projects'.
- Each iteration cycle involves a set of activities associated with five technically-focused *workflows* (requirements, analysis, design, implementation, test). The extent to which the activities concerned with each workflow are involved will be different for each phase.

Figure 13.11 is an attempt to model this fairly complex process. In particular, the workflow balance associated with each phase has been shown as a histogram (the relative heights of the bars should not be taken too literally; these are meant to be *indicative*). We begin with a brief look at the roles of each phase, and then discuss the activities of the workflows. Some aspects are omitted: the *iterations* are really a project-specific element; and we concentrate on design-related activities, which means that we say little about the implementation and testing workflows.

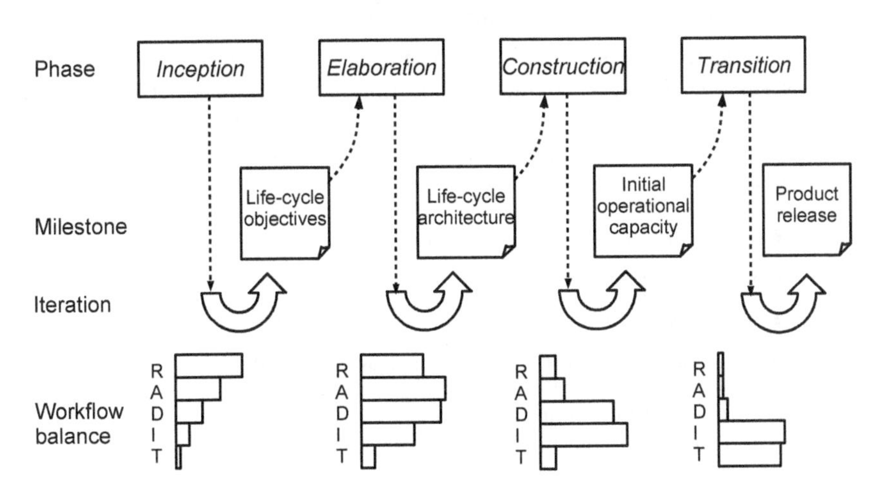

FIGURE 13.11: Organisation of the Unified Process

The UP phases

The phases are very project-driven and create a framework that emphasises the strong user links and iterative practices that characterise the UP (and that characterise the agile forms that we examine in the next chapter).

1. **Inception.** This phase is primarily concerned with project planning activities. However, we might note that establishing the feasibility of a project may lead to the creation of one or more *exploratory prototypes*. The milestone for this phase, *life cycle objectives* is largely concerned with the eventual deliverables for the phase (documents, models, prototypes), but does include a *candidate architecture*, representing the initial architectural design for the eventual application.

   The milestone documents should also include:

   - an initial domain-based class model;
   - a set of domain-based use cases;
   - initial analysis and design models.

   Although iteration is implicit, the normal expectation is that this phase should require only a single cycle. We might also note the use of *use cases*, a very distinctive feature of the UP and one that has been elaborated in later developments.

2. **Elaboration.** From a design perspective, this is a much more significant phase. Its purpose is to create the *architectural baseline* (top-level design) that will underpin further development. In doing so, there is the expectation of producing further use cases, as well as addressing questions about quality factors, including those related to performance.

   The milestone for this phase, the *life cycle architecture* is a partial working system. The UP documentation does emphasise that this is not an exploratory prototype, although it can be argued that it is fairly close to being an evolutionary prototype. The resulting set of models will include:

   - static domain-based class models;
   - a fuller set of use cases;
   - analysis models (both static and dynamic);
   - static and dynamic architectural design models.

   The emphasis upon model-building and the need to ensure consistency between them, means that some degree of iteration is likely to be needed.

3. **Construction.** Despite the name, this phase still involves some design activity. The *initial operational capacity* milestone corresponds to the delivery of a beta version of the application. Hence its goals include:

   - completion of any requirements elicitation or analysis tasks;
   - completion of models.

   Clearly, the detailed physical design tasks form an essential element of this.

4. **Transition.** The purpose of this phase is to lead to the final milestone of *product release.* Hence it is unlikely to lead to any design activities unless a need for these was revealed when exercising the beta version produced in the preceding phase.

The UP workflows

A characteristic of plan-driven design methods that may by now be fairly obvious is that terms such as 'construction' or 'implementation' can mean quite different things when used by methodologists, and in particular, may involve significant design activities. For that reason we will look at all five workflows, although only with regard to the design issues that are involved in each one.

1. **Requirements workflow.** This workflow relies extensively on the use of *use case modelling.* Not only is the use case a rather distinctive characteristic of the UP, but a use case also has the advantage of being able to record both functional and non-functional attributes. The primary role of a use case diagram is to identify the boundaries of a use case, as shown in the example of Figure 13.12. The detailed specification of a use case is not a part of the UML model, and designers often use text-based templates, although other forms such as message sequence diagrams can also be used (Ratcliffe & Budgen 2001, Ratcliffe & Budgen 2005). Most books on the UML do discuss this issue, and there is a good introduction to use case modelling in (Arlow & Neustadt 2005).

   One of the benefits of employing use cases is that they provide a good mechanism for verification of a design model against requirements. A scenario derived from a use case can provide a walk-through mechanism that directly links the two stages of development. Use cases also provide a framework for the analysis workflow.

2. **Analysis workflow.** The UP interprets 'analysis' in the conventional 'black box' sense of producing a model that describes *what* an application is to do, but not *how* it will be done. As always, the distinction between analysis and design is not completely clear-cut. The objectives of this workflow are to produce *analysis classes* that model elements (objects) in the problem domain and to generate *use case realisations* that show how the analysis classes interact in order to create the behaviour required for a given use case.

   As always, identifying the analysis classes remains a challenging problem. In the case of the UP this is supported through the use cases, as each use case can be separately analysed to identify both the analysis classes and the way that they collaborate. The task of identifying the classes still remains though, and relatively long-established techniques

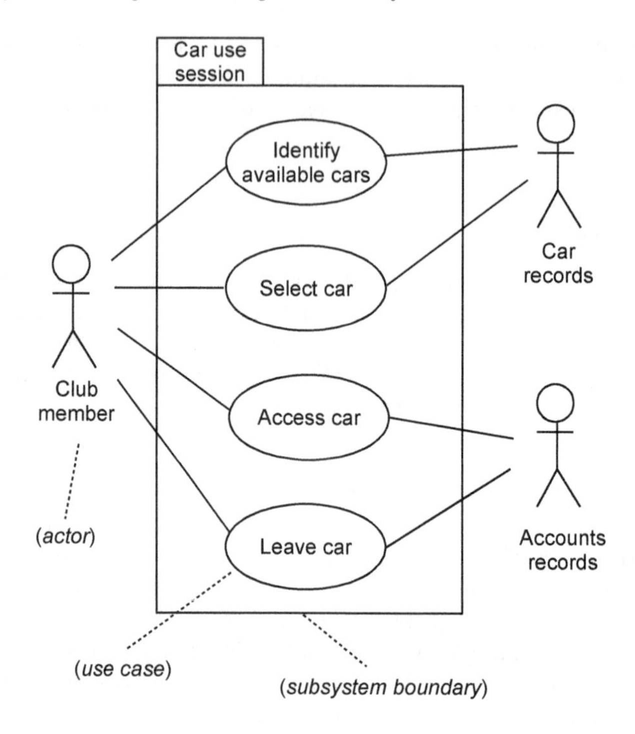

FIGURE 13.12: Simple example of a use case diagram

such as *noun-verb analysis* and CRC brainstorming (*class-responsibility-collaborator* is described in (Beck & Cunningham 1989)) can be used.

The output from this workflow can be modelled by using a range of UML notational forms, including:

- *class diagrams* to model the static properties of the relationships between objects (uses, inheritance);
- *collaboration diagrams* and *message sequence diagrams* to model the dynamic aspects of the relationships between objects.

For a given use case, collaboration diagrams provide a static model of the relationships between objects—describing *which* elements are involved, and *how*—rather than the *when* aspects (the actual interaction) that are described in forms such as message sequence diagrams. In particular, an interaction can only occur between objects within the context of a collaboration. Collaboration diagrams can be used to model class collaborations (the *descriptor* form) and object collaborations (the *instance* form). An example of a simple use case from the CCC, described using a common tabular form, and the associated collaboration diagram is shown in Figure 13.13 (some detail has been omitted for clarity).

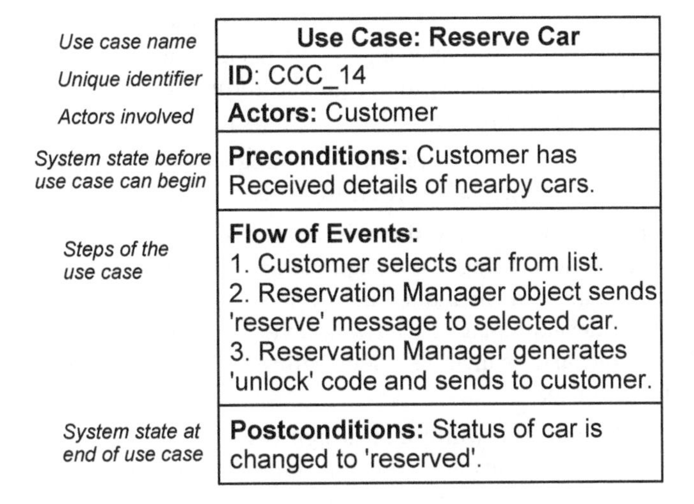

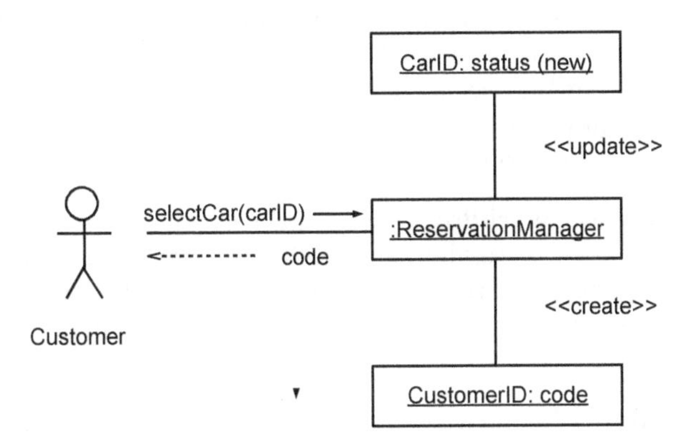

FIGURE 13.13: A UML use case and associated collaboration diagram

The UML *package* notation can be used to group analysis classes and use case realisations as *analysis packages*. Since packages can also contain other packages, the analysis model can itself be considered as being a package. A key role of this notation is to help keep track of the analysis elements, with the overall analysis architecture being defined by the high-level analysis packages.

3. **Design workflow** This forms a major element of both the elaboration and construction phases and is intended to provide the white box description of how the black box model from the analysis workflow is to be achieved.

   Many of the activities are similar to those described for Fusion: identifying design objects; when to create objects from the static classes; and the persistence, or lifetime, of objects. Another consideration for this workflow is the more structural question of how to employ the inheritance mechanism. And once again, the package mechanism can be used to manage the outcomes, although the term *subsystem* is now used in place of *package* to indicate that more detail is provided.

Use cases are expected to provide a linking thread between analysis and design and may themselves need to be elaborated further.

The UP design workflow therefore addresses many detailed issues concerned with developing the design classes and their interfaces. Perhaps the main change in the modelling involved, apart from elaboration of detail, is the development of *behavioural* viewpoint models, largely through UML statecharts, which also helps address the issues of object creation and persistence.

As with Fusion, the white box design tasks of the UP are largely concerned with elaborating detail and ensuring consistency between the different parts of the design model. Most of the key architectural decisions have already been made as part of the analysis workflow, although obviously this can be revised by the activities of the design workflow if necessary. In that sense, like Fusion, the key structural decisions are made early in the process reflecting the more compositional approaches employed by these object-oriented methods.

4. **Implementation workflow.** This is primarily concerned with producing executable code, with the main design elements being confined to the decisions involved in translating the design model to actual implementation. Overall, this phase is not expected to form an active part of the design task.

5. **Test workflow.** As might be expected, this involves no explicit design activities beyond any required in response to the testing outputs.

The UP—some observations

The UP design process has a much more complex structure than the other methods examined in this chapter, including Fusion. The interweaving of development phases and workflows produces a much more complex process than the more linear forms of the other design methods we have examined, and as such forms a much more challenging management task.

A key element of the UP, and one that particularly distinguishes it, is the *use case*. This concept provides a valuable framework that provides manageable-sized analysis tasks; a thread through requirements, analysis, and design; and a mechanism that not only partitions the tasks for these workflows, but also provides the means of being able to cross-check for consistency and correctness. In many ways the more complex UP process model would probably be impractical without the unifying theme of the use case.

13.7 Empirical knowledge related to plan-driven design

The use of plan-driven design methods is perhaps more associated with the ideas of *advocacy* than those of rigorous empirical assessment. In the early days at least, design methodologists tended to be codifying their experiences (a quite reasonable approach to knowledge transfer in the circumstances) rather than exploring their limitations. There are a number of possible explanations for the lack of useful empirical knowledge about design methods, including the following.

- By their nature, design methods both evolve and are intended to be adapted to meet different needs. This means that any attempt at evaluation of the processes involved usually lacks a widely-established and well-defined 'baseline design method' to study.
- Software development (at least before the emergence of open source software) was largely conducted by commercial organisations—and knowledge about how it was done may well have been considered as being of too much value to allow for open publication.
- As we have observed, the development of ideas about design methods rather 'tailed off' in the 1990s. However, this was also the period in which interest in empirical research about software engineering was beginning to become more established, and so researchers were perhaps less likely to view design methods as a topic of interest.

The most appropriate form of empirical study to use in studying plan-driven design development is probably the *case study*, since the subject matter needs field studies with relatively long-term data collection, and there are also likely to be many variables of interest. Unfortunately, these are challenging to perform for such a topic, and examples do appear to be lacking in the literature. Nor do there seem to be any useful observation studies.

An alternative form for collecting information about experience with design processes is the *survey*, and while these may collect data that can be considered as more 'shallow' for these purposes, a survey does offer scope to collect and aggregate user experiences. Such a survey was conducted by Edwards, Thompson & Smith (1989) to look at the use of SSADM, and other surveys have been conducted to address questions about object-oriented methods. In the rest of this section we examine their findings and observations about the use of plan-driven approaches (there were other findings, but here we concentrate on the methodological aspects).

Knowledge about SSADM. The survey by Edwards et al. was related to Version 3 of SSADM, and was conducted on an organisational basis. Requests were sent to 310 organisations in industry and in local and central government organisations, from which they obtained 117 responses, with 72 (23%) of these being usable. (Given that anything above 10% is regarded as being a good response rate for a survey, the researchers did well.) At the time when they conducted their survey, projects tended to be using programming languages such as COBOL and many were also concerned with database management systems. For the responses, team sizes involved ranged from teams with fewer than 10 up to ones with over 200 developers.

The survey revealed that SSADM was rarely used in full, with developers being selective about when to make use of it. In particular, it was noted that, while responses tended to be positive:

- the *techniques* were generally found to be effective, but were time-consuming to employ;
- the *entity-life-history* diagram presented most problems for developers when modelling and was the form of notation most frequently omitted;
- the *physical* design step was the most challenging, perhaps in part because the method provided little in the way of detailed guidance about how to perform this for a particular platform.

Knowledge about OO design. The survey conducted by Johnson & Hardgrave (1999) used two groups of participants: experienced developers and trainees. Also, its focus was upon their attitudes and preferences with respect to object-oriented methods and tools, rather than upon the forms of the specific methods.

Separate survey forms were used for the two groups, and the sample included 102 experienced designers and 58 trainees. Since the survey was conducted on-line, there are some methodological issues regarding sampling and representativeness, as the authors do acknowledge.

The authors also observe that the degree of comparison (between methods) that could be achieved was limited. This was chiefly because "a theory explaining attitudes and behaviour toward, and use of OOAD methods does not currently exist". Hence they argue that the survey offered an inductive approach that would "use facts to develop general conclusions" and lay some of the groundwork for the development of such a theory.

In terms of comparisons, the survey was chiefly concerned with the degree of training provided for developers, their familiarity with, and preference between, different object-oriented methods, and their attitudes towards them.

The methods covered by the survey were chiefly what we might term as 'first generation' design methods, although the set of methods did include *Fusion* as well as the three methods that were subsequently brought together in the *Unified Process*. Their findings included the following.

- A relatively large proportion of time was spent on analysis and design of objects, relative to the time spent on coding and testing when compared with 'normal' expectations for software development. It was thought that this might be because creating the design for an object was more time-consuming than the equivalent task for other architectural styles.
- Adopting object-oriented methods involved a steep learning curve (although the respondents did regard this overhead as useful once the knowledge had been acquired). This is consistent with an earlier study on the overheads of learning analysis and design for different forms of software architecture (Vessey & Conger 1994), which also observed that OO concepts could be challenging to use.
- The limited expectations that respondents showed about code *reuse*. Here the authors observed that "one of the most advertised benefits of OO is reuse, yet developers rated this lower than trainees". Their conclusion was that "this is perhaps a case of trainees believing the hype about reuse and developers actually realising the difficulty in creating reusable code".

Overall, the survey did find high levels of satisfaction with object-oriented methods, both for analysis and design. However, as the authors did caution, this might have partly been an artifact caused by the self-selection of participants, since those with a positive view of OO might be more likely to have responded to the invitation to participate.

Unfortunately, there are no later surveys that can be readily identified in the literature, probably because the use of plan-driven methods with OO effectively 'ran out of steam' with the development of the Unified Process. The systematic mapping study described in Bailey et al. (2007) found relatively limited empirical research into the object paradigm as a whole, with most emphasis being upon studies involving metrics,

---

Key take-home points about plan-driven design practices

Knowledge transfer. Plan-driven design practices provide *method knowledge* as the means of organising the design process, with this acting as a substitute for domain knowledge where appropriate. The process itself is usually organised as a sequence of 'analysis and design' steps that provide a solution modelling process for an ISP that is more akin to that for a WSP.

Design input. Although plan-driven forms provide a structure that can help organise the design process, it is important to appreciate that the design decisions made within this are still ones that are based upon the needs of the application. Using a plan-driven approach requires that the designer possesses method knowledge, an understanding of the problem, and an appreciation of relevant design criteria.

Plan-driven design strategies. These usually consist of a mix of a *representation* part; a *process* part; and a set of *heuristics* used to help adapt a strategy or method to a particular type of problem. The process part incorporates *declarative knowledge* about how the design task should be organised as a set of steps. The heuristics provide *procedural knowledge* about how to employ the method for a particular type of problem.

Context. Plan-driven methods implicitly assume a 'waterfall-like' development context. This is chiefly because they build their initial analysis models around the elements of the requirements specification. While this doesn't preclude making later changes to the model, the assumption is that the requirements specification is a comprehensive description of the design goals.

Architectural form. Earlier plan-driven methods were concerned with relatively simple architectural styles such as call-and-return, where the dependencies (*coupling*) between design elements used only a limited set of forms. Use of plan-driven approaches has proved more challenging with such styles as the object-oriented model, where there are many forms of interaction that can occur between the elements of a design.

Effectiveness. There are few empirical studies available to provide any insight into this. One benefit of using a systematic plan-driven approach is that it helps constrain the structure of the resulting design model, which in turn may assist with later evolution of the model. Perhaps one of the main disadvantages of the plan-driven approach is that it can be bureaucratic, with a significant overhead of documentation—although as in the case of SSADM, that may be considered as a benefit.

Chapter 14

Incremental Design in Agile Software Development

In the previous chapter we examined the use of plan-driven design approaches and noted some of the limitations of this form—in particular, the need to have available a fairly complete specification of the functional and non-functional requirements for intended application at the beginning of the design process. Given that much of software development is, and always has been, concerned with developing new and innovative applications, the difficulty of being able to specify a complete set of requirements at the outset of a project is self-evident.

In practice, for most applications, the evolution of requirements and the development of a design tend to be intertwined and to play complementary roles in enhancing understanding just what the application needs to do. A recognition of this interplay therefore led to various efforts being made to find ways to 'systematise' this process.

In this chapter, we first look at some early ideas about how to support this interplay, and then at the ideas that came together under the *Agile Manifesto*. Not all of the ideas discussed, particularly in the first sections, are usually considered as being 'agile'. However, they most definitely are not plan-driven either, and so fit well with the ideas of this chapter. As far as possible, we focus upon the specific issues related to software *design*, but, as we noted with the Unified Process, when we move away from 'waterfall thinking' about the role of design, it becomes much harder to compartmentalise its contribution to the process as a whole.

14.1 Using software prototypes

As observed above, a major limitation of a linear development process such as a plan-driven one, is the need to identify exactly what is needed from an application at the very beginning of the process. There are many reasons why this is largely impractical: those commissioning the application (who may be the developers of course) do not have a comprehensive picture of what it will do; the picture that they do have may need to change between starting and finishing development because of a growing understanding; and there may be many uncertainties and possible omissions in the understanding of what is required.

In other engineering domains, such a situation is often resolved by the construction of some form of *prototype* which can be used to investigate both what is needed, and how it might be delivered. The same idea can be used with software, although the ease with which software can be changed can make its role and use rather different. For more traditional branches of engineering, a prototype is usually regarded as a 'first of kind', or takes the form of a model that can be realised on a (reduced) scale. In automotive engineering, a prototype of a car is definitely a first of kind (and sometimes the last too), while scale modelling may well be used with complex civil engineering projects. (As an example, the engineers constructing the Thames Barrier down-river from London built a scale model of their intended structures in the nearby marshes.)

Analogous roles for prototypes are not easily found in software development, where there is no manufacturing cycle and no real equivalent to scale modelling. Indeed, for software it is quite possible that the prototype will actually become the product in due course, which would make no sense for our examples from other branches of engineering. However, while the *form* of a prototype might be different for software, the *reasons* for creating prototypes are very similar to those found in other disciplines. Prototypes are constructed in order to explore ideas about the requirements and design of an application more completely than would be possible by other means (such as modelling).

A useful categorisation of major prototyping forms as used with software was produced by Floyd (1984), and this recognised three principal roles for a prototype.

1. *Evolutionary.* This is the form that is closest to the idea of *incremental development* of an application. The software is changed gradually by changing the requirements specification as experience is gained with

using the application, and changing the application to fit those changes. In this form, prototyping is used to develop a product and the prototype gradually evolves into the 'end' product.

One benefit of this strategy is that it allows a product to be made available quickly where the market or other factors mean that this is needed urgently. It can then be refined through subsequent releases (Cusumano & Selby 1989).

2. *Experimental.* In this role, the purpose of the prototype is to assist with evaluating a possible design model for all or part of an application, by developing a 'quick build' in advance of a more complete design and implementation. The prototype can be used to assess such aspects as a user interface or an algorithm, and may well be built using tools that produce an implementation that is quite different to the form intended. Such a prototype is definitely intended as a throw-away item.

3. *Exploratory.* Such a prototype is usually used to help clarify user *requirements* and perhaps to investigate how using the application might change the way that users work within an organisation. In this role, it can act as a form of partial feasibility study, focusing on the issues of interest alone, and again, such a prototype is really a throw-away item.

Whatever the purpose, a prototype can provide a useful intermediate step towards addressing an ISP. The *experimental* and *exploratory* forms can be viewed as simplifying the elements of the problem (whereas a plan-driven approach seeks to simplify the process), while the *evolutionary* approach structures the way that the interactions in Figure 1.1 are managed. This is illustrated in Figure 14.1.

The idea of prototyping, and of evolutionary prototyping in particular, fits in well with the concept of an *emergent* organisation. An emergent organisation is one that is "in a state of continual process change, never arriving, always in transition" (Truex, Baskerville & Klein 1999). Where the software it uses needs to reflect this characteristic, then prototyping can help explore what the needs of an application are as well as how it can be created.

That all said, the adoption of a prototyping approach does not offer any particular guidance about how the development of a design should occur. What it does imply, is that the design process does not take place in a single step, but is organised on an *incremental* basis.

From a designer's perspective, prototyping does not necessarily involve generating code. This is particularly the case for human-computer interaction (HCI) elements of a system (sometimes referred to as *interaction design*) (Sharp, Preece & Rogers 2019), where mock-ups of screens might be employed to help gauge responses from potential users.

Within design more generally, a non-code form of prototyping that is often useful is the *story-board.* Story-boards can be used for eliciting requirements

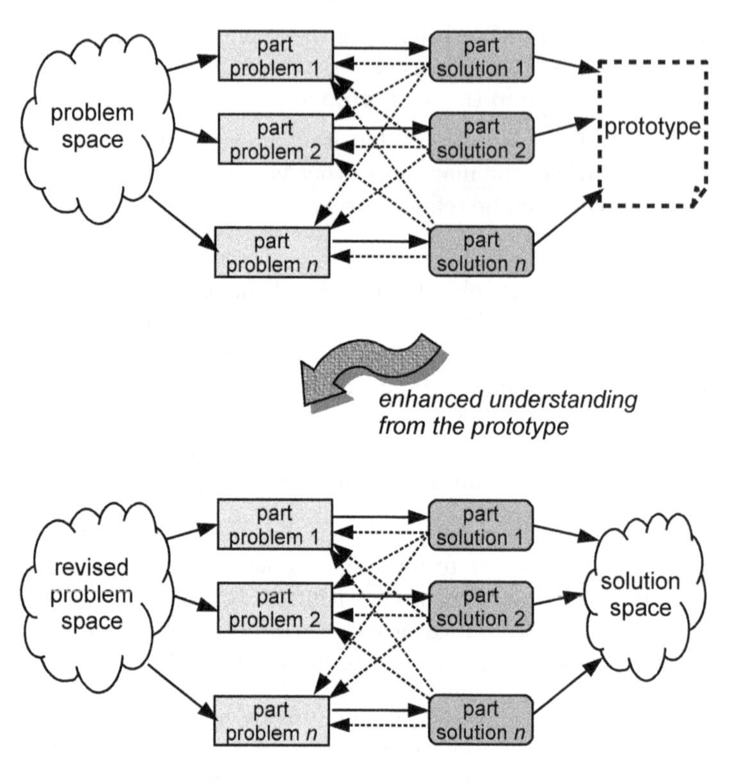

FIGURE 14.1: Use of an evolutionary prototype

as well as for design, and provide a pictorial overview of a process. Story-boards can provide useful support for *design reviews*. These, conducted both with other team members as well as with end-users and customers, can form a valuable element of incremental design, as they can convey system behaviour without needing to go into technical detail. (And with the benefit that a story-board does not easily become an evolutionary prototype!). The story-board can also be used in conjunction with the idea of the *use case* discussed in the context of the Unified Process, both for developing and also for recording the details of a use case. Figure 14.2 shows a simple example of the use of a story-board when developing the CCC software. In practice this would probably be sketched if it is being used within a design team, although it may well be turned into a more elegant presentation for a customer.

Experts prototype concepts. PvdH #55

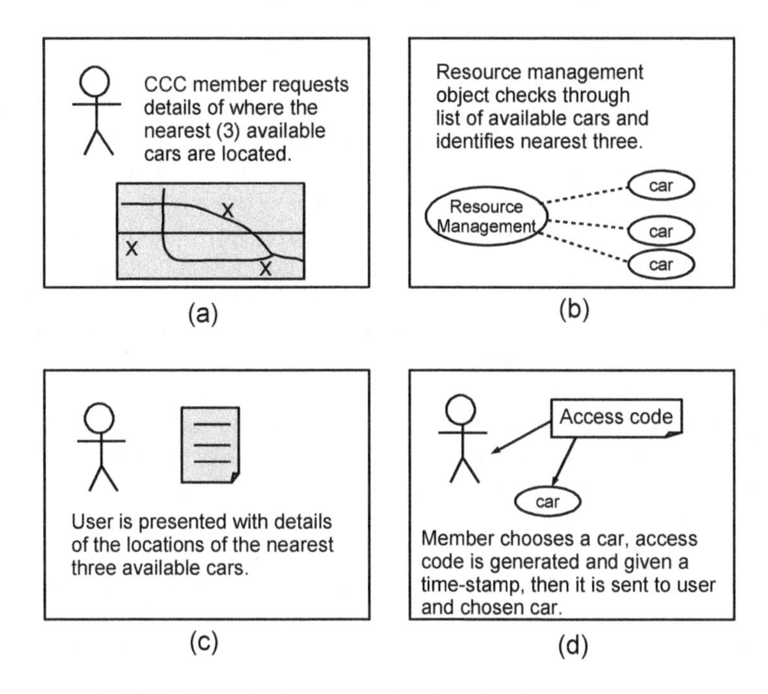

FIGURE 14.2: An example of a simple story-board

14.2 Incremental development and the spiral model

Incremental development can be viewed as offering a way of making the design process more flexible than plan-driven approaches, which achieve their goals chiefly by turning the process into a highly structured procedure.

Indeed, while plan-driven forms seek to 'organise' the design of solutions to ill-structured problems so that the process is closer to the sort of process used for well-structured problems, the incremental forms described in the rest of this chapter take a rather different approach. Incremental strategies use a repeated application of the form modelled in Figure 14.1. This accepts that we are seeking a solution to an ISP, but breaks up the process of developing the solution into 'chunks', with each chunk being completed in a separate *phase*. As with plan-driven forms, the goal of doing so is to simplify the design process, but now the design process is essentially opportunistic and needs to address a number of viewpoints in each increment.

The way that the 'chunking' is organised differs for each of the approaches that we examine in the following sections. In DSDM the stages are organised around development activities, whereas for XP they are organised (differently) around the functionality of the application. The main issue is that by progressing in this way, the development team can periodically re-assess the ISP being tackled based on understanding gained in the preceding stages.

From a design perspective, this represents something of the inevitable trade-off. Plan-driven forms tend to become overly complex when used with other than relatively simple architectural styles. Equally, many key decisions are made early in the development process, when the nature of the intended application is still incompletely understood. However, plan-driven strategies do provide a (relatively clear) path to the end goals. Incremental design provides greater flexibility, but some decisions still need to be made early on (such as architectural form) and it may be harder to keep a sense of direction towards a solution. This may be more challenging for larger projects (Boehm 2002, Elbanna & Sarker 2016), and there are questions of the skills needed for designing in such a context and the effectiveness of agile development for knowledge transfer about design skills (Jacobson, Spence & Seidewitz 2016).

One of the earliest attempts at breaking away from the perceived rigidity of plan-driven forms was the *spiral model* devised by Barry Boehm (1988). In many ways, Boehm's spiral model was quite 'visionary', and it explicitly assumed the use of prototyping in an incremental process by which the design model was developed through a set of incremental stages. At each stage in development the next set of developments for the design model were planned in terms of:

- the *objectives* for that particular stage;
- the *options* and *constraints* that were to be explored;
- evaluation of the *risks* involved in choosing between the options;
- formulating a *plan* for the stage, which might involve developing a new prototype or involve other activities, according to the conclusions of the risk analysis.

Within the context of a prototyping strategy, perhaps the most important element to highlight from this list of actions is that of *risk assessment*, since this links back to the very reason why such an approach might be needed. In particular, given that one of the risks implicit in the use of prototyping is that the sense of 'direction' might become lost, the spiral model offered a mechanism for monitoring and controlling the development process.

Incremental design is an approach that is less commonly encountered in other forms of engineering design. As a technique it is unlikely to find much favour with civil engineers ("after we have built the central span of the bridge, we'll work out how to construct the approaches"), any more than with other engineering disciplines such as chemical or electrical engineering. This is not to argue that it is not a valid engineering approach, given sufficient planning and control, but rather that it is an approach that only makes sense when used with a medium such as software where there is no manufacturing stage, and within the type of dynamically changing context often met in software-based business use and in the emergent organisations that are made possible through the use of software applications. It is also widely employed for development of open source software (OSS) as discussed in Chapter 18.

> *Experts design throughout the creation of software. PvdH #22*

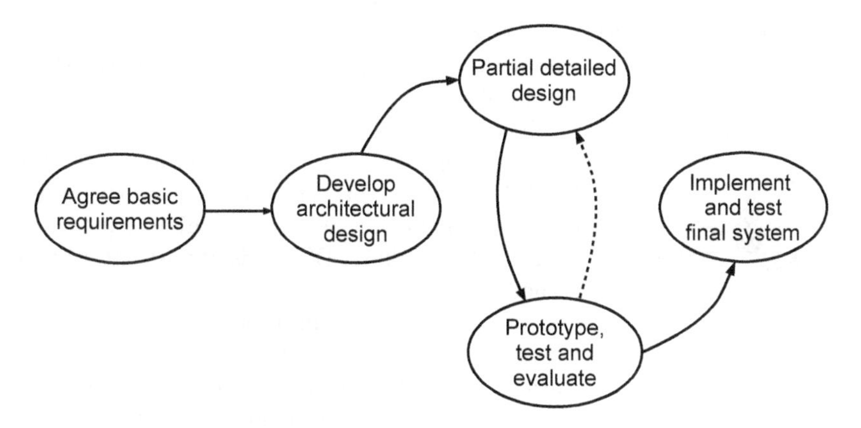

FIGURE 14.3: Profile of an incremental development process

One issue that can easily become blurred where incremental development is employed is the distinction between design and implementation. An incremental approach is likely to involve some interleaving of design stages with implementation and evaluation. However, the design stages are likely to be more concerned with 'detailed design' issues, occurring within a context established by a set of architectural design decisions made early on. Figure 14.3 illustrates this point schematically, and in the next section we will see that our example of DSDM broadly conforms to this form of structure.

Valuable and influential though the spiral model was, in itself it didn't offer any very strong procedural or declarative guidance about how development, and design in particular, should be organised as a set of 'chunks'. This was addressed in the later development of thinking about *rapid application development* or RAD.

14.3 RAD: the DSDM method

RAD, or *rapid application development* can be considered as describing a process of development that is essentially a structured form of incremental development. RAD has attractions for developing commercial applications and was widely popularised by the many works from James Martin. Approaches based upon RAD can be viewed as fitting in between plan-driven methods and more lightweight agile forms.

Our example of a RAD development process is the *dynamic systems development method*, generally known as DSDM. This originated in 1994, managed by the 'not-for-profit' DSDM Consortium (the associated trademark being formed by reversing the second 'D' character). In 2016 this became the 'Agile Business Consortium', a change which highlights the direction of travel that DSDM represented in terms of the evolution of RAD. The main textbook available that addresses DSDM is (Stapleton 2002), although the consortium has produced quite extensive documentation in the form of on-line material. This documentation also links DSDM to other approaches such as the UP and agile development practices (as we might expect).

Rapid Application Development

Strictly speaking, DSDM is almost entirely concerned with managing and controlling the RAD process and makes no assumptions about design strategy, architecture or any of the other factors that characterised plan-driven methods. This is not to say that these issues are ignored in the DSDM process, simply that DSDM practices make no assumptions about their particular form. (Indeed, it is possible to employ DSDM for non-software development projects, although obviously we will not pursue that aspect here!)

Throughout the DSDM process, there are two distinctive elements that emerge very strongly:

- the roles, responsibilities and activities that *people* perform;
- the effects of *business* needs upon design decisions.

Since we can regard DSDM as representing a transition stage in thinking about software development practice (rather than providing specific thinking about *design* itself), we will examine its features fairly briefly, although again, we will put most of our emphasis upon those aspects that can be considered as being related to 'analysis and design'. We will first look at the *principles* that underpin DSDM, since these provide the rationale for its particular structure and practices. After that we will examine how the principles are interpreted within the DSDM development cycle.

14.3.1 The DSDM principles

There are nine principles, which in many ways look forward to the *Agile Manifesto* that we describe in the next section.

1. *Active user involvement is imperative.* DSDM identifies a number of such roles such as the *ambassador*, who "guides the developers in their activities to ensure that the solution being developed will accurately meet the needs of the business"; and the *advisor*, who is tasked with representing a particular user view.

2. *DSDM teams must be empowered to make decisions.* This is a corollary to the first principle, since active user involvement is incompatible with the need to obtain decisions from higher-level project management.

3. *Focus is on frequent delivery of products.* DSDM favours managing increments on the basis of the 'product' (including documentation) over increments of the 'activities' themselves, with an emphasis upon allocating fixed periods of time for performing various activities.

4. *Fitness for business purpose is the essential criterion for acceptance of deliverables.* DSDM takes the view that delivery within the required time period is of paramount importance, and that once the functionality has been established, the structure of the software can be re-engineered as necessary. (This can be seen as down-playing the role of design.)

5. *Iterative and incremental development is necessary to converge on an accurate business solution.* The key issue here is that iteration needs to be part of the organisation's procurement strategy, with the aim of achieving continuous improvement in the software through its use.

6. *All changes during development are reversible.* DSDM accepts that the possible need to backtrack to an earlier state (of the design model) is inherent in incremental design. Hence configuration control is an essential and all-pervasive element of the development context.

7. *Requirements are baselined at a high level.* Unlike plan-driven forms, DSDM is mainly concerned with freezing requirements at a high level (which in turn relates to the need to choose an architecture), and allowing more detailed aspects to evolve as necessary.

8. *Testing is integrated throughout the life-cycle.* This is a major departure from plan-driven thinking, which tends to consider testing as a post-implementation element, but again looks forward to ideas about agile development.

9. *A collaborative and co-operative approach between all stakeholders is essential.* Involving the stakeholders in the development process implies the need for 'lightweight' change control procedures, allowing for short-term redirection of a project through consensus.

A distinctive feature of the philosophy, and one that strongly influences the process, is that of *time*. In a plan-driven context, the functionality that an

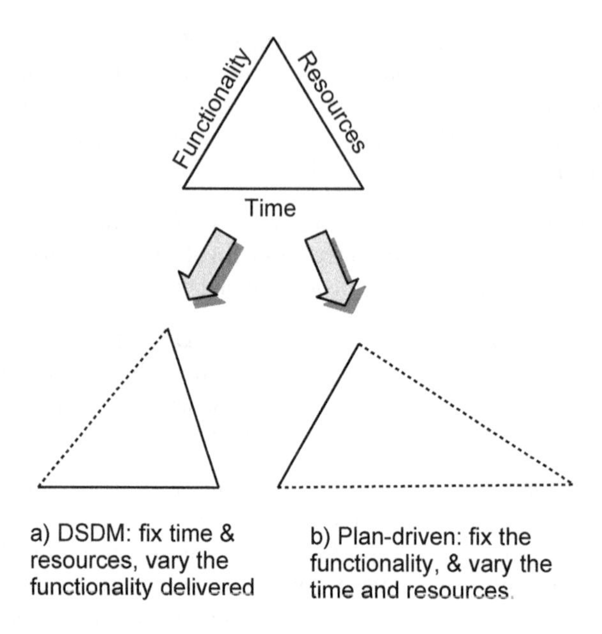

FIGURE 14.4: Varying the design constraints (time, functionality and resources)

application is expected to provide, embodied in the requirements specification, is fixed, with the time and resources needed to achieve this being varied as necessary. But in the DSDM context this is inverted, time and resources are considered as fixed during a cycle, and the deliverable functionality is allowed to change, as illustrated schematically in Figure 14.4.

Obviously this philosophy is one that rather constrains the scope for using DSDM. No-one would advocate its use for developing avionics software ("we didn't have time to complete the module that allows the landing gear to be lowered, but go ahead with the test flight anyway"). On the other hand, where being first to market is important, some limitations in an application might be more acceptable if agreed by developer and end-user.

An important concept in DSDM is that of *timeboxing*. This is a process "by which defined objectives are reached at a pre-determined and immovable date through continuous prioritisation and flexing of requirements using the *MoSCoW* rules" (Stapleton 1999). (Note that timeboxing is a process, rather than some form of object, although the DSDM literature does then refer to individual 'timeboxes'—and it is difficult not to do so!) A timebox is normally expected to last between two and four weeks. The MoSCoW rules used to prioritise requirements are summarised in Table 14.1. In essence the role of timeboxing is to ensure that a project team has a clear focus and deadline at any point in time and to manage the fixed-time aspect by providing guidance on the acceptable degree of functionality that an increment should have. In planning a timebox the recommendation is that about 60% of effort should be directed towards implemented 'must have' requirements, with the remaining 40% split between 'should' and 'could' haves. This means that if it proves impossible to achieve all of the goals within a timebox, there should still be enough resource and time to ensure that the 'must have' elements are delivered.

TABLE 14.1: The MoSCoW Rules for prioritisation

Rule	**Interpretation**
Must have	Those requirements that are fundamental to the system, such that it will be inoperable if they are not met.
Should have	Important requirements that would be mandatory if more time were available, but for which some sort of work-around can be achieved in the short term.
Could have	Requirements that can safely be left out of the current deliverable, being desirable but not essential.
Want to have but won't have this time	Requirements that are on the waiting list for the future.

Of course, assigning requirements to categories can be quite a difficult process, and this is where the emphasis that DSDM places upon a collaborative approach to development is important. Team members who insist that all of their requirements are clearly in the 'must have' category are unlikely to make useful contributions to this process!

14.3.2 The DSDM process

The DSDM process is generally described as being a *framework* made up from five *phases*, and provides little that can be considered as prescriptive. Figure 14.5 provides a simple model of this. In many ways this is an elaboration of the model shown in Figure 14.3 for incremental development, although the feedback involved in iteration is rather more complex. The description here focuses chiefly on the impact that this model has upon the design process.

Feasibility study. This phase is meant to be kept short, and should deliver a *feasibility report*, together with an *outline plan* for development that provides more detail. One option is to develop a *feasibility prototype* that can be viewed as largely exploratory (and of course, need not actually involve writing software if a form such as story-boarding is used). From a design perspective this is likely to explore possible architectural forms, rather than provide anything more detailed.

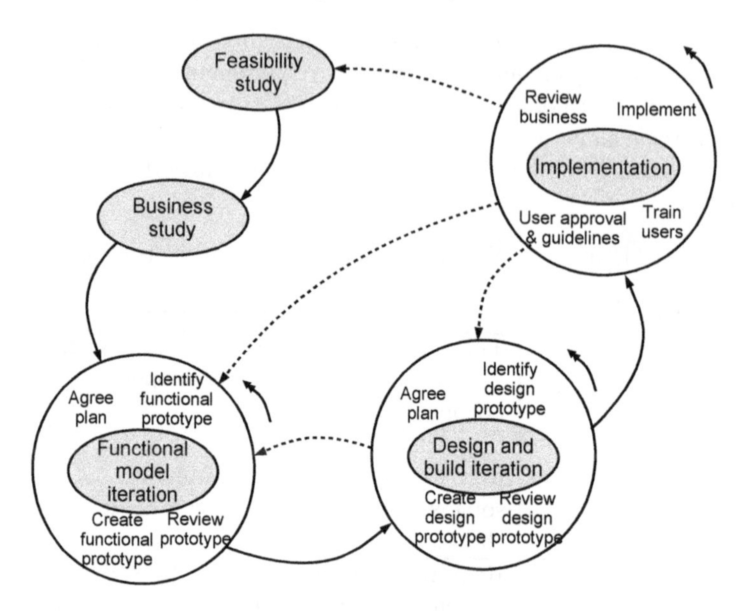

FIGURE 14.5: The DSDM development process

Business study. This again is intended to be of short duration and to be highly collaborative. It leads to what we can consider as being the first real design decisions, embodied in the *system architecture definition*. This identifies both the architectural style to be employed and also the main elements of the system architecture.

Functional model iteration. This is really the element that concentrates on black box modelling of what the application is to do within the given business context. So this phase does emphasise modelling of business needs, without going into detail about such non-functional aspects as security and performance. The main outputs are models such as class models and data models and *functional prototypes* that are expected to be more or less complete in terms of their user interface, along with some basic degree of functionality.

Design and build iteration. This is the phase where the detailed decisions about implementation are made. The main output produced is still classified as being a prototype although it may well incorporate the key functionality of the application. This phase produces a mix of white box models and actual code and incorporates a strong element of testing.

Implementation. The emphasis here is less upon developing the software as getting it into productive use, including collaboratively addressing any needs for training material and documentation.

So, what are the distinctive features of DSDM, viewing it particularly as a exemplar of RAD practices? These include an emphasis upon management and control (essential in any form of incremental development); strong emphasis upon user involvement throughout the process; fixing of delivery time and varying of deliverable functionality (timeboxing); the non-prescriptive view of application architectural style; and the extensive use of prototyping.

As an incremental process, DSDM organises the 'chunking' of the development (and design) process around fairly classical lines of black box modelling followed by white box modelling. And like the agile approaches that we examine in the rest of this chapter, DSDM links its procedures closely into the business processes related to an application. In contrast, for plan-driven forms, the influence of the business aspect is largely confined to requirements elicitation and system modelling, with later stages being much more concerned about software structures.

14.4 The agile manifesto

Obviously the trend towards more iterative and flexible approaches to development didn't end with RAD, and a number of other incremental development practices were emerging through the late nineties. In 2001 this led an influential group of 17 methodologists to agree upon what has come to be termed the *Agile Manifesto*. In essence, this states a set of *values* that they believe are embedded in 'agile software development'. The manifesto is shown in Figure 14.6.

Agile development

In many ways the manifesto was a culmination of various attempts to move to more flexible ways of developing software, not the beginning of it. Indeed, as we noted earlier, Barry Boehm (2002) argued that neither agile nor plan-driven approaches have all of the answers. Indeed, agile development still needs structures if it is to avoid becoming a 'code and fix' approach and in the next two sections we examine two quite different examples of agile forms (and many would consider DSDM to be an agile development form too).

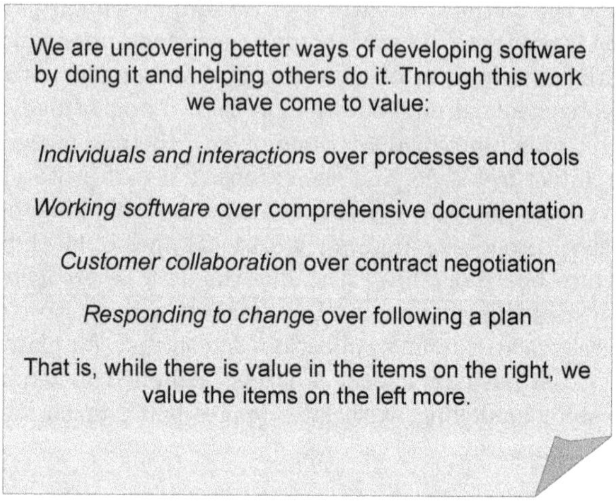

We are uncovering better ways of developing software by doing it and helping others do it. Through this work we have come to value:

Individuals and interactions over processes and tools

Working software over comprehensive documentation

Customer collaboration over contract negotiation

Responding to change over following a plan

That is, while there is value in the items on the right, we value the items on the left more.

FIGURE 14.6: The Agile Manifesto

14.5 Extreme programming (XP)

Extreme programming, commonly abbreviated to XP, was devised by Kent Beck, one of the leading agile methodologists. He described XP as "a light-weight methodology for small-to-medium-sized teams developing software in the face of vague or rapidly changing requirements" (Beck 2000). And his argument for including 'extreme' in the name was that "XP takes commonsense principles and practices to extreme levels".

Like the other approaches in this chapter, it relies upon an evolutionary and incremental design process rather than a fixed set of development activities. Indeed, part of the flexibility of agile approaches is that they can adapt to the needs of the problem, allowing the development team to focus on the key issues, whether these be architectural design, testing etc. And as Boehm (2002) has observed, agile methods such as XP do involve quite a lot of planning (and re-planning), although the emphasis for agile tends to lie in the planning itself rather than the generation of documented plans.

XP is quite well documented, perhaps in part because it has a stronger 'technical' element than many other agile methods that tend to put greater emphasis upon managing processes. One thing is certain though, it does not lend itself to being described using the same framework that was employed for describing plan-driven methods (representation, process, heuristics)!

XP is usually characterised as being based upon five *values* and twelve *practices* that are underpinned by the values. The values themselves are fairly high-level concepts: *communication*; *feedback*; *simplicity*; *courage* and *respect*.

Indeed, just the names of these indicate the strong people-flavoured aspect of agile. We observed in Part I that the idea of software engineering as a 'social discipline' only emerged as the discipline grew in maturity. Indeed, it can be argued that agile methods take a much more rounded view of software engineering and software development than plan-driven approaches because of this recognition that development is indeed a social process as well as a technical one.

To understand their role, it is probably best to look at the twelve practices, and how they are influenced by the values. Once again though, we focus attention mainly upon the ways that the practices influence the role of software design in XP.

1. **The planning game.** Like other agile methods, XP is iterative, and at the start of each iteration the team and the customer undertake a 'planning game', seeking to capture *user stories* on a set of 'story cards'. There are some obvious similarities to use cases of course.

   The basic idea is that the customer provides a set of user stories that represent a piece of application functionality (so really, these are smaller than a use case). Ideally a user story can be captured in a few sentences, along with any non-functional implications, and can then form the basis for resource estimation (people, time, money). Each story also forms the basis for an associated test set (more about this later).

2. **Small releases.** The XP philosophy is to produce frequent releases, with each forming a small advance on previous ones—usually advancing by implementing more user stories.

3. **Metaphor.** This is an important design aspect, mapping user stories on to the classes and objects needed to achieve the required functionality.

4. **Simple design.** The philosophy here is clear. A design should avoid unnecessary 'bells & whistles', and when considering changing any design element or creating new ones, the question to ask should be "does the customer really need this feature?".

5. **Test-first programming.** The (unit) tests should be written *before* writing the actual code and testing should be a continuous process.

6. **Refactoring.** This is a reflective element that involves restructuring the code without changing its functionality. The aim is to simplify the code so that as the design evolves it is kept as simple as possible.

7. **Pair programming.** All code is written by two programmers who work as a team at a single computer, discussing their work as they go.

8. **Continuous integration.** Code is integrated into the evolving application as often as possible, and after passing unit tests.

9. **Collective ownership.** The code of the application is owned by all, and any member of the team may make changes to it when they feel it necessary.

10. **On-site customer.** The customer works *with* the team to answer questions, perform acceptance tests, and monitor progress.

11. **40-hour week.** Iterations should be organised so that overtime is not needed, on the basis that tired developers make mistakes.

12. **Open workspace.** Developers share workspace and use shared coding standards, following clearly-defined conventions for such aspects as identifiers, layout etc.

Interesting as these are (and there are few that would be likely to appear in a plan-driven method), as observed above, our main concern here is how these influence the process of *design*. Indeed, one thing we might note is that there are no particular assumptions in the practices about architectural style or the detailed form that the implementation will take.

So, based upon the values and practices described above, we can identify a number of characteristics that can be considered to form the design strategy of XP. As might be expected these are much more concerned with principles that design activities should follow, rather than the exact form that the activity should take. Indeed, as one measure of the nature of 'agility', it is impractical to try to draw a process model for XP.

- *The KISS principle.* The idea of 'keep it simple stupid' is to remind designers to avoid over-complicating the design model. Implicitly of course, complicated structures create technical debt and make future changes more difficult. In an agile context, where those future changes are happening as part of the development process, this is particularly important. So a simple design minimises the set of classes and methods, avoids duplication of logic, and avoids unnecessary items–while of course, meeting the criterion of 'fitness for purpose'.

- *Use CRC cards.* We encountered the use of class-responsibility-collaborator cards in the previous chapter when looking at the Unified Process. They are a valuable way of documenting a design, providing a place to note design issues, and also helping to identify where an element is becoming overly complicated.

- *Use 'spike' solutions to reduce risk.* A *spike solution* can be regarded as a form of exploratory prototype, and is used where the team need to decide how to proceed with a particular element of the system that is proving to be complex. By building a "quick and narrow implementation" (Pfleeger & Atlee 2010) the team can obtain valuable feedback about the design options.

- *Have a metaphor.* This is really a description of the application that the project aims to produce. In some ways it is similar to the idea of the *architectural pattern* that we discuss in the next chapter, providing a simple message that describes the overarching form of the application and maps its key features on to software structures.
- *Refactor.* After adding a factor to the code, the team should look to see if this can be simplified. This complements the goal of keeping the design as simple as possible (the KISS principle).

An important point here is that while the XP literature tends to refer to 'code', there is absolutely no reason why the design should not be modelled with diagrams when the team thinks this will help. The emphasis in XP upon developing through a series of small increments does not present an obstacle to doing this either, provided we remember that sketches are probably the appropriate form to use.

Viewed as an incremental development process, XP 'chunks' the increments based on application functionality rather than on performing specific development activities (in principle, each iteration may involve combinations of all of the 'classical' activities).

14.6 Agile development: Scrum

Our second example provides a rather different example of agile thinking, with much greater focus upon the processes rather than the technical issues of producing a design.

The basic ideas underpinning *Scrum* date from the mid-1990s. (The name isn't an acronym, and refers to the way that team members in a game of rugby huddle together when they are trying to advance.) Two key references are (Schwaber & Beedle 2002) and (Schwaber 2004)[1]. The original motivation appears to be the question that Ken Schwaber asked about software development: "why do the defined processes of the CMM (Capability Maturity Model) not measurably deliver?". Without digressing too much from our focus on design, we should note that the CMM is primarily concerned with assessing how dependable an organisation's development processes are—with a strong emphasis upon the idea of *learning* from previous projects. Schwaber recognised that development processes were dealing with ISPs (whereas the CMM implicitly treats them as WSPs by assuming that such experiences are transferable) and that developers needed to accept change rather than expect predictability.

[1]A short definitive (downloadable) guide from the developers of Scrum is available at https://www.scrum.org/resources/scrum-guide

Scrum is largely focused on the development process rather than on particular practices (in contrast to XP) and aims to provide:

- an agile process to manage and control development;
- a wrapper for existing engineering practices;
- a team-based approach to incremental development;
- control for the chaos of conflicting interests and needs.

It has some similarities with DSDM and in particular, it extends the idea of *time-boxing* and uses it to control almost all of the activities.

As a result, the process is iterative and incremental, and is centred upon time-boxed development cycles termed *sprints*. It begins with a vision of the system, a baseline plan of likely cost, and some time-frames, with the vision possibly stated in market terms rather than technical (product) ones. An important concept in Scrum is the *product backlog*. This is a list of the functional and non-functional requirements that are needed to deliver the vision. The product backlog is:

- prioritised to ensure that high-value elements get a high priority;
- divided into a set of proposed releases (with each release being generated from a sprint);
- changed to reflect the way that the business needs evolve.

Creating and managing the backlog is a major task in Scrum.

As noted above, the *time-box* is a major tool for Scrum and time-boxes are used to organise most of the major activities of Scrum. We examine some key ones here.

Release planning meeting. Effectively, a *release* is a deliverable increment of an application. The planning meeting seeks to establish a plan and goals for the development team and the rest of the organisation, which may well be based upon user stories. It may involve some element of design activity concerned with architectural issues and their consequences. The resulting *release plan* establishes the goal of a release, identifies priorities, risks and the functionality intended for the release.

Sprints. Each *sprint* creates a new internal increment of the product, and when enough increments have been created and accumulated for the product to be of value, it is released. A sprint usually involves an iteration that occurs within a time-box lasting *one month* (obviously this can be varied, but if so, it is likely to be less than one month rather than more). If necessary, to maintain the time-box the team may decide to reduce the functionality or scope of the product. Equally, if things are going well, they may be able to implement additional elements from the product backlog. The sprint is the core organisational element of Scrum.

Sprint planning meetings. Assuming that the sprint will last for one month, the time-box allocated for this meeting is *eight hours*. Part of this time is spent on prioritising the product backlog, and this is then used to craft a *sprint goal*. Again, we might expect that this will involve some thinking about design, but given time constraints this is unlikely to produce a detailed design model in any form (although it might extend an existing one).

Sprint reviews. This has a time-box of *four hours* (assuming a one-month sprint). Among other things it provides input to the next sprint planning meeting and (possibly) considers questions about design choices.

Sprint retrospectives. This is a meeting allocated a *three-hour* time-box aimed at reviewing the process of the preceding sprint and looking for possible actionable improvements.

The daily scrum. A team meeting with a *fifteen-minute* time-box that reviews what has been accomplished since the last meeting and plans for the day's work.

Clearly, the management and control of time-boxing is a really important element of Scrum, and much of the guidance available tends to be focused upon this and upon the different roles that team members may have.

Scrum can be considered as being *architecture-agnostic*. The process is completely separated from the software architecture (in complete contrast to plan-driven forms) and should be usable with a wide range of architectural styles. Indeed, it simply assumes that the team have the necessary experience to produce the necessary design ideas, implicitly by some opportunistic process. We say a little more about this in the next section.

Finally, there is the question of how the incremental process is 'chunked'. For Scrum this is a combination of application functionality and time. Technical factors have little or no influence on the way that sprints are organised.

14.7 Refactoring

Refactoring is the process of modifying the internal structure of source code with the aim of improving its organisation, while preserving the external behaviour. There may be various reasons why the code needs to be improved, but refactoring is something that should help reduce technical debt, aid with identifying reusable elements and generally improve such aspects as readability, coupling and cohesion.

The need for refactoring is apt to arise from the piecemeal manner in which a design may evolve using agile practices such as XP and Scrum. It is also important that refactoring does not introduce defects into the code, and this is where the emphasis upon testing that is often found in agile practices is important. The code to be refactored should have passed all of the necessary tests before it is refactored, and should pass them again afterwards. (There is of course no reason in principle why refactoring should not be undertaken with designs produced from plan-driven methods, but these tend not to place such emphasis upon the early deployment of unit testing regimes.)

In the classic work on refactoring, Fowler (1999), the author observes that this is the opposite of what we conventionally expect to do. Normal practice is to produce a design model and then turn this into code, but with refactoring, the aim is to take bad design or coding, and rework it to produce a good design. So now, rather than developing our understanding through entwined evolution of requirements and design, we are looking at refining a design through entwined design and code.

Refactoring is really an essential adjunct to the use of agile strategies. During agile development, the design model evolves slowly (and opportunistically) through the lifetime of the project, rather than being developed before implementation begins. And if it is to evolve successfully, the design elements need to be structured in such a way that they too can evolve and be changed.

On the positive side, refactoring does also provide the opportunity to make good use of relationships such as inheritance and mechanisms such as polymorphism. We observed in the previous chapter that these presented challenges for plan-driven object-oriented methods, and we will see in the next chapter that design patterns tend to make limited use of these forms. And of course, these are very code-oriented mechanisms, and ones that are not easily incorporated into design models. Using a more reflective approach by refactoring a design may therefore offer a way to make good use of them.

So, what sort of activities does refactoring involve? Here we very briefly look at what is involved in a few of these. For more details about these and other forms, see Fowler (1999).

Extract method. This involves decomposing long and complex methods by identifying blocks of code within them that can be used to form the basis of an independent method, which ideally, might then be reusable.

Move method. This is where a method is encapsulated as part of an object but it would be more appropriate for it to be part of another object, perhaps because its main task involves modifying the variables of this second object.

Replace array with object. This typically occurs where the elements of an array may have different roles and purposes (which is rather contradictory to the idea of how we use an array, but is how we may end up using one). Essentially what we create is a data object with a set of methods for manipulating the different elements (rather like the 'car object' proposed for the CCC).

Refactoring may be performed for other reasons of course, with performance being one example of a possible motivation. Identifying candidate classes for reorganisation may also be based upon the use of code metrics, and upon consideration of such issues as *code smells*, a topic that we address in the next chapter.

Of course, refactoring also requires time and effort, which may be difficult to spare in a project, particularly as refactoring is something that we perform with code that is working correctly. Here again, we encounter the trade-off between time spent on performing restructuring in order to make future development activities easier, and the need for a team to deliver code. In a survey by (Tempero, Gorschek & Angelis 2017) examining what might impede developers from refactoring, the authors observe that "the decision to refactor is ultimately a business decision", reinforcing the view that refactoring is really not a design issue alone.

So, from the perspective of the software designer, refactoring can be viewed as 'design adjustment' rather than 'design modification'. It involves reorganising the design elements, while retaining the functionality and behaviour of the design. So it can be regarded as purely affecting the design from the constructional viewpoint with the goal of improving largely non-functional aspects of the design model.

14.8 Empirical knowledge about design in agile development

This chapter (and the previous one) cover some quite different forms of development life-cycle and there is an interesting review and summary of many of these in Ruparelia (2010) (an article that is described as a 'history column').

While there is relatively little empirical knowledge for plan-driven forms, the agile concept emerged at more or less the same time that empirical studies were becoming firmly established in software engineering. As a result, there is what might sometimes seem to be an endless stream of evaluation frameworks for comparing methods and of empirical studies related to various aspects of their use, adoption, adaptation, etc. This section is not a systematic review of knowledge about agile development (Kitchenham et al. 2015); rather it is a selection of a few studies that can be thought of as examples, or that particularly address issues highlighted in the previous sections.

So, here we concentrate on evaluations that involve some element of assessment of how far a method provides guidance about *design*. Because they are better defined and quite widely used, this is also largely structured around the evaluation of DSDM, XP and Scrum. Indeed, it is worth observing that while none of the approaches covered in this chapter provides much detailed guidance about 'design as a verb', the amount of guidance provided about the design process can be considered to taper off as we go through the methods, with Scrum providing little or no guidance at all about how a design might be developed.

14.8.1 Empirical knowledge about DSDM

There appear to be few studies that focus upon DSDM although it does appear in a number of studies that compare different agile methods in some way, such as that by Qumer & Henderson-Sellers (2008). In particular, there appears to have been little research addressing design explicitly, although a useful survey to examine the effectiveness of user participation in DSDM was performed by Barrow & Mayhew (2000). In this, the authors observed that "There was total agreement among respondents that consensus between participant stakeholders was the basis of participation in the approach, and the likelihood of such consensus being possible through co-operation and collaboration between stakeholders during development was deemed to be very high (in fact there were no views to the contrary)".

14.8.2 Empirical knowledge about agile methods

Since many empirical studies compare and contrast between different agile methods, this section covers both XP and Scrum.

An early systematic review of studies of agile methods by Dybå & Dingsøyr (2008) noted that at that point in time, most of the primary studies available were of XP. While providing a useful review of the benefits and limitations of agile methods, the topics covered in the primary studies were largely focused on themes related to social and organisational aspects, or on the adoption of agile practices.

Later systematic reviews seem to have mainly focused on answering questions about specific aspects of using agile methods, but these do not seem to have investigated much about how design is managed (beyond looking at quite specific forms such as user-centred design). The mapping study by Diebold & Dahlem (2014) examined use of agile methods in industry, finding that (like the findings from the empirical study of SSADM discussed in the previous chapter) few agile methods are used in their entirety. Their study examined 68 projects and analysed these for their use of 18 agile practices that were largely derived from DSDM, XP and Scrum. They also looked at whether a practice was used 'out of the box' or adapted in some way. The six practices that were used most widely were: time boxing; planning meeting; learning loop; evolving and hierarchical specification; daily discussion; and product vision. All were essentially 'management' aspects, which reflects the emphasis of agile development.

Given a lack of specific studies of design, it is helpful to look at studies that have investigated the risks implicit in using agile forms to see how far design issues arise there. An interesting study by Elbanna & Sarker (2016) surveyed the adoption of agile practices in some 28 organisations spanning a wide range of business sectors. The risk associated with development and deployment that was mentioned most often (23 organisations) was the accumulation of *technical debt*. The authors observe that "debt can quickly accumulate owing to the need to significantly reduce development time, adhere to strict time boxing, and constantly deliver functional requirements for business use". As a result, software becomes more complex and difficult to maintain. Nearly as important was the separation of development and IT teams within an organisation (21 organisations). Not only are working practices for the two types of team very different, but the need for the project team's product to work in a specific operational environment forms an additional constraint, which may not be adequately recognised.

14.8.3 Empirical knowledge about refactoring

The role and importance of *refactoring* has led to a number of empirical studies. The study by Tempero et al. (2017) was mentioned in Section 14.7, and examined factors that might impede the process of refactoring.

The study by Chen, Xiao, Wang, Osterweil & Li (2016) provides the results from a survey of agile software developers with 105 respondents who had a wide range of roles. The results showed diverse views about the tasks involved and planning for refactoring, influenced by a respondent's roles and also the type of applications being developed. In particular it was noted that while teams do careful planning for many aspects of agile development, "planning for refactoring is not done nearly so carefully".

Key take-home points about designing in an agile context

Incremental and agile development practices are largely concerned with how the process of software development is organised and hence provide relatively few lessons about design explicitly.

Evolution of the design model. In contrast to plan-driven forms, incremental and agile development practices and methods do not try to simplify the process of 'solving' an ISP by organising this as though it were a WSP. Instead, they sub-divide ('chunk') the process into a series of steps with the aim of simplifying and structuring the design process through the use of smaller and less complex goals for each increment. The way in which the sub-division into incremental steps is organised forms a key characteristic of a particular method.

- *DSDM* organises the increments around fairly a classical development model (black box model leading to white box model) and the use of time-boxing within that.
- *XP* organises the increments around fairly small additions to application functionality.
- *Scrum* uses a mix of application functionality bounded by strict time-boxing.

None of them is concerned with specific architectural styles and the associated practices are largely concerned with team management.

Refactoring. This involves reorganising the *constructional* aspects of the design model while retaining the behavioural and functional aspects. It can be viewed as providing a mechanism that helps refine the effects of piece-meal evolution of the design model.

Transfer of design knowledge. None of the agile methods examined in this chapter provided any specific mechanisms by which experience could be transferred between experienced and less experienced designers. However, this might be misleading, since agile forms place a high value upon mentoring, as well as the use of team meetings and the sharing of information.

Chapter 15

Designing with Patterns

So far, in this third part of the book, we have been largely concerned with design approaches that involve following procedures (methods) to create applications. In this chapter we consider a quite different way of producing a design (or part of a design), by reusing abstract ideas about design structures that have been found to work well by others. In a way, it makes use of forms that could be considered as being more related to 'design as a noun' in order to support the activities of 'design as a verb'!

Reuse of experience comes in various forms, and the idea of designing an application by reusing some form of *pattern* has attracted a lot of attention since the 1990s. Patterns offer another way of codifying experience about design, and of transferring it to others—although there are those who consider that they provide a knowledge transfer mechanism that is more appropriate to experienced designers than beginners.

The pattern concept can be employed at different levels of abstraction, and with different architectural styles, and although it is usually associated with object-oriented development, it is by no means restricted to being used with such architectures. However, as with all mechanisms for transferring design knowledge, patterns provide no automatic guarantee of success, and do require care in use.

15.1 Patterns as a mechanism for knowledge transfer

The concept of the *design pattern* is one that originated in another discipline, and is rooted in the work of the architect Christopher Alexander (Alexander et al. 1977). He has described it in the following words.

> "*Each pattern describes a problem which occurs over and over again in our environment, and then describes the core of the solution to that problem, in such a way that you can use this solution a million times over, without ever doing it the same way twice.*"

As a simple example, we could have a pattern for building a school. The problem that it addresses is a generic one (to support the process of educating children and young adults), and the format of teachers and classes is more or less universal. However support for this might be realised in many different, but related, ways, with the differences between them reflecting such issues as educational structures, age group, climate etc.

While the 'school pattern' describes how something might be *structured*, there are other familiar patterns that describe how things *behave*. Bus services provide an example of a behavioural pattern with easily-recognised components (bus stations, stops, routes, timetables etc.). Catching a bus in an unfamiliar place might involve local variations, such as where to board, and when and how to pay, but the idea is a familiar one that usually requires little explanation.

Used in the context of *design*, a pattern can provide a generic solution to some problem which recurs in various forms, and which itself may be sub-part of a larger problem. A pattern can help a designer gain understanding of the characteristics of a particular problem as well as providing a strategy for addressing it. If it is well documented, it should also provide some ideas about the context where it might arise, as well as of any possible consequences (usually in the form of technical debt) that might be incurred when using that pattern.

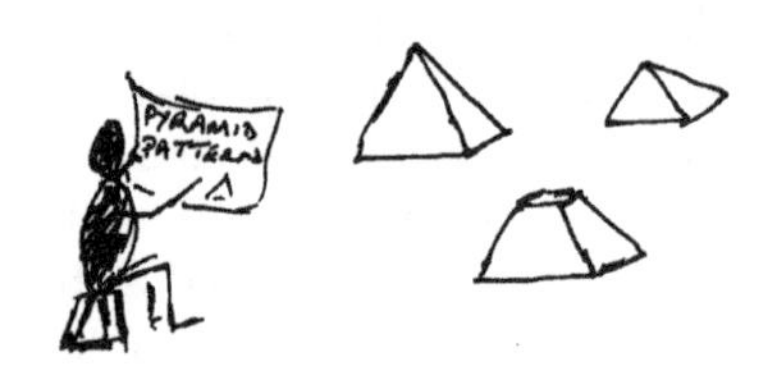

Build-a-pyramid pattern

In some ways, the idea of the pattern comes much closer to embodying the traditional master/apprentice model for transfer of knowledge and expertise than forms such as plan-driven design methods can hope to achieve. Although design methods also address ways of finding solutions, patterns educate about problems too, and the recognition of these is an important part of learning about design, and one that is not always easy to achieve without some form of guidance. Patterns also provide scope for providing peer-to-peer exchange of design knowledge in manageable 'chunks', in a form that matches the observed practices of designers. One of the characteristics of expert designer behaviour that was observed in the pioneering study by Adelson & Soloway (1985) was the employment of 'labels for plans', whereby a designer would recognise and 'label' a sub-problem that they knew how to address, leaving them free to concentrate on the less familiar aspects of a problem. Implicitly, they intended to reuse a 'pattern' that they had employed previously. In the context of design patterns, this labelling also has a role, by providing a useful *vocabulary* that allows designers to share ideas.

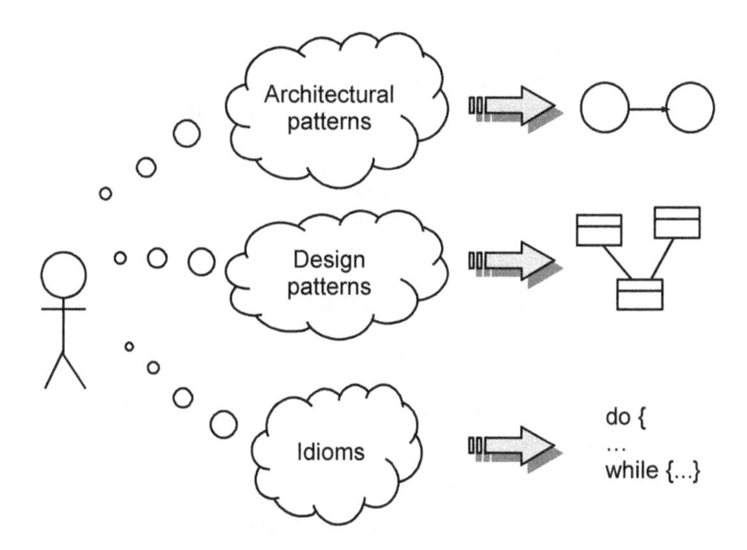

FIGURE 15.1: Patterns at different levels of abstraction

The design pattern concept is not restricted in its use to 'detailed design' activities. At the programming level it is generally referred to as an *idiom* (Coplien 1997) and at an architectural level, the concept of the *architectural pattern* is well established (Buschmann et al. 1996, Bass et al. 2013). While we will not examine idioms in any detail in this chapter, we will examine the forms and uses of both 'design' and 'architectural' patterns. Figure 15.1 illustrates their roles.

It is important to appreciate that a pattern is not a reusable piece of software. (That would be termed a *framework* (Fayad & Schmidt 1997).) Rather, a pattern is a form of conceptual *knowledge schema* (Détienne 2002), forming a piece of generalised design knowledge that has been extrapolated from experience. In particular, it describes how all or part of a design solution (model) is organised. In practice, patterns are:

- more concerned with forms of coupling between elements that are based on using *composition* rather than *inheritance*;

- represent a way of sharing the collective experiences of skilled designers (knowledge transfer);
- are categorised in terms of some abstract (and recurring) design issue, providing a description (pattern) for how this might be addressed.

So, in terms of the way that we address an ISP, how does the use of patterns influence the overall design process? Basically, the use of patterns can reduce the opportunistic nature of the interactions that occur during design. Patterns impose an element of structure upon the design model which might relate to behavioural or constructional constraints, as illustrated in Figure 15.2. (We may also recognise patterns in the problem space which will influence how we approach the design process, but this aspect is omitted from the diagram.)

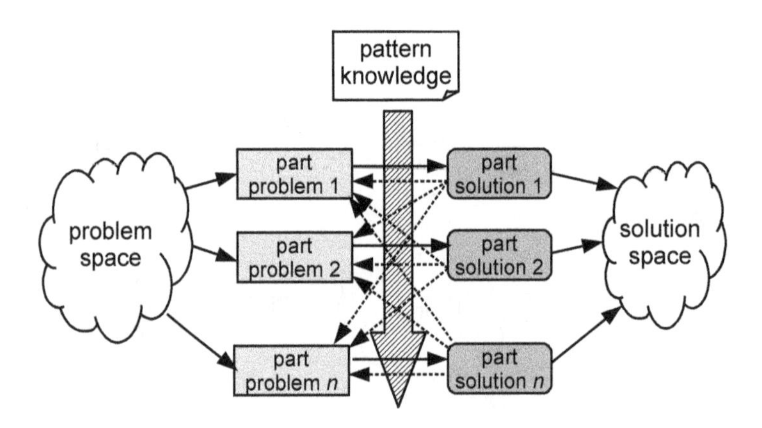

FIGURE 15.2: How patterns influence 'solving' an ISP

To end this introductory section, we look at a simple example from the CCC, not with the aim of providing specific solutions at this point, but rather with the aim of recognising that building an application that manages its processes may well draw upon experience with other applications.

Patterns in the CCC

The organisation and operation of the CCC has some characteristic 'patterns' that we can see in many other applications. Some examples are listed below.

- The core business of the CCC involves managing resources that are used by members, who reserve and book cars. This is characteristic of very many other applications such as car rental; the traditional 'lending library' providing short-term access to books, music etc.; theatre bookings; and fishing clubs. All of these manage some form of 'resource' that is used by clients.
- The billing process is common to very many applications, based on membership details, car use and member 'credit'.
- The reservation process is likewise familiar in many ways. Booking theatre seats, train journeys and the like, may involve selecting preferences from a fixed set of choices about time, seating etc.

So, we may already have some ideas about how these aspects could be managed in the CCC application.

15.2 Architectural patterns

Looking at the examples from the CCC at the end of the preceding section, one of the patterns that we could recognise in this is what could be termed the 'renting a resource' pattern. While this is concerned with the *function* of the application (and patterns occur in requirements specifications too of course), it describes a high level pattern that is concerned with the overall *behaviour* of the application.

When we turn to the role of design in meeting such a need, the most abstract form of pattern that we encounter is the *architectural pattern* that provides the outline form of the overall system architecture (Buschmann et al. 1996, Bass et al. 2013). In Chapter 6 we briefly examined these from the perspective of *architecture*, while here we consider their role as *patterns*. The main concern of such a pattern is to describe the overall *organisation* of a software application—in particular, how the sub-systems should be structured and what the relationships between these should be. 'Organisation' in this context may relate both to static issues of configuration, as well as to run-time behaviour for the interactions between the elements.

As with other patterns, the basic description of an architectural pattern needs to address the following three factors.

- The *context* within which some recurring situation occurs. Examples of this might be the need to decouple elements of an application, or to incorporate the ability to dynamically update information sources while a system is operating.
- The *problem*, which is related to the issues that the context may present for software development.
- A *solution* that addresses the needs of the problem in a suitably general way. This may be described in terms of the type of element making up the application, the forms of interaction between elements, the topology of the elements, and the likely constraints and trade-offs involved.

Architectural patterns have some synergy with ideas about *architectural style* although they are not necessarily restricted to being realised using a particular style. In principle at least, any architectural style can be employed with a pattern providing that the *form* of the elements can provide the necessary structures and that their interactions can be organised to fit the pattern.

Perhaps because of the large-scale perspective involved, the set of architectural patterns is much smaller than that of design patterns discussed in the next section. However, as with all patterns (and styles) they do provide a useful vocabulary for the designer who is wishing to exchange ideas or explore options. In the rest of this section we briefly examine three examples of architectural patterns. (A much more detailed and analytical explanation of a useful catalogue of architectural patterns is provided in (Bass et al. 2013).)

15.2.1 Model-view-controller (MVC)

The model-view-controller pattern describes a widely-used form of organisation that is employed for many applications that involve some degree of interaction with end-users. Here, the context is one of needing to decouple some form of information store (the 'model') from the way that it is shown to different users and viewed on different devices (the 'views'). The problem is how to achieve this decoupling while also keeping the application responsive to user actions. To achieve this, *MVC* divides such an application into three types of element.

- The *model* incorporates the core functionality and associated data for the application.
- A *view* provides information about the model to the user, presented in a particular way. There may well be multiple views, corresponding to the needs of different types of user, or to the need to present information in different ways. It also interprets and communicates user actions to the controller.

- *Controllers* handle user input. Each view has an associated controller that responds to any choices made through the use of relevant forms of input (keyboard, mouse, touchscreen,...).

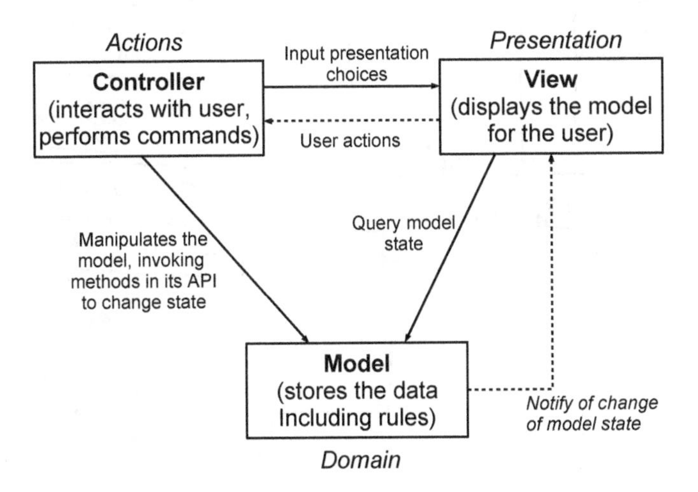

FIGURE 15.3: The model-view-controller pattern

Figure 15.3 shows this schematically (note that there are many forms of visual description for MVC). Note that both views and controllers send requests to the model, but that the model does not send out any requests, and it only provides responses through the view. Requests sent by a controller do not create a direct response, instead they cause the model to be modified, and the corresponding view to be notified that the state of the model has been modified. The application may also directly change the model (with no user interaction) of course, and again, when this occurs the views are notified. The view can then request details of the revised model (or part-model) and display it to the user in response to their input (the mechanism for doing this is provided by the *Observer* design pattern described in the next section.

The important characteristic of MVC is the decoupling of the internal model from the associated interactions with the user (*separation of concerns*). We can recognise this pattern as describing many familiar but different, interactive applications. Word processors, spreadsheets, web browsers etc. all fit this model. Decoupling the 'knowledge store' from the way that it is presented, as well as from the different forms of interaction, makes it possible to implement the application so that it can be used on different operating systems or with a range of devices.

Figure 15.4 shows a sketch of how MVC might be used with the CCC application. This is really a rather 'behavioural' sketch, where the designer is thinking out how the MVC model will handle the situation where a customer makes a booking. The booking results in a modification to the record set for

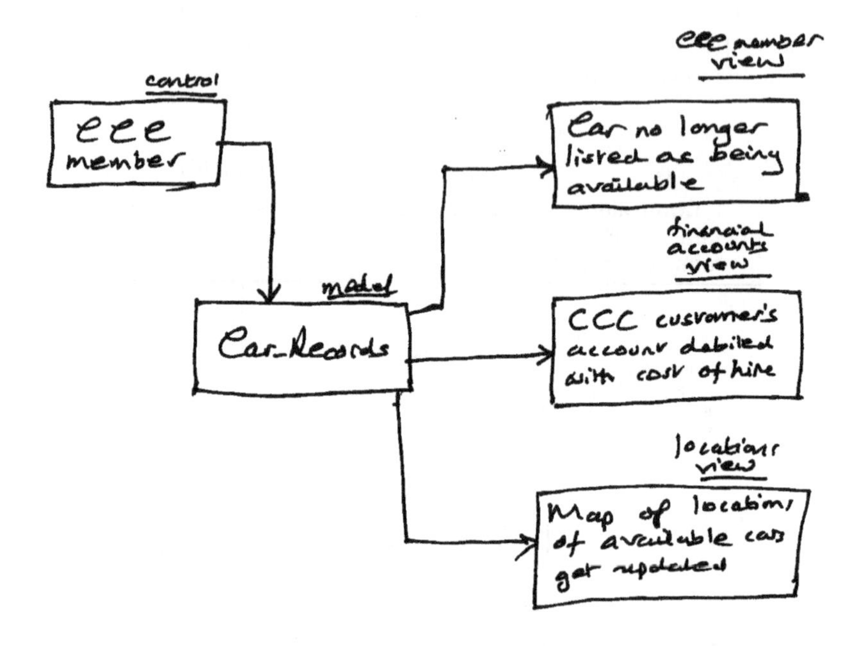

FIGURE 15.4: An MVC interpretation of the CCC application

the cars (the chosen car is no longer available to other customers), and it has some consequent effects upon the different views (such as the symbol for that car on the city map changing colour).

15.2.2 Layers

The *Layers* pattern addresses a context where different elements of the software need to be developed and to evolve separately, requiring that interaction between them should be kept to a minimum. This means that the resulting software is likely to be portable and can readily be modified to meet new needs. Essentially, Layers represents a very stratified approach to addressing the issue of 'separation of concerns'. When using Layers, the solution adopted is to group those elements that provide a particular service into a 'layer', with the layers being organised in a hierarchy, so that the layers providing higher levels of abstraction depend on lower ones, but not vice-versa.

While Layers may not be used as universally as MVC, it does provide the structuring for some very important software applications, with networking and operating systems being key examples of where this pattern is particularly useful. Figure 15.5 illustrates the use of Layers to implement the OSI model used in computer networking.

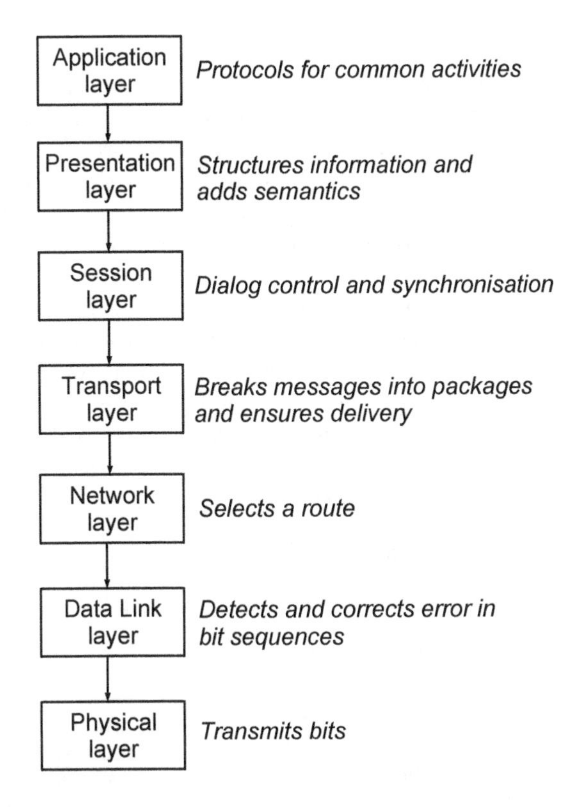

FIGURE 15.5: A layers example: open systems interconnection model

The OSI model is probably quite an extreme example in terms of the number of layers employed. At a minimum, there need only be two layers. For *Layers* to be used appropriately, it is necessary that:

- any dependencies are unidirectional (each layer depends only upon the layer 'below');
- a layer will normally only use the layer immediately below it (exceptions are possible but are better avoided);
- a layer should embody some particular role or functionality, and ideally making modifications or adaptations to the application should only affect one layer.

Implicitly too, the use of Layers may impose a run-time overhead, since there may be many method (sub-program) calls involved in communicating between the upper layers and the bottom-most layer, which is where the required action is eventually performed.

15.2.3 Broker

In everyday life, the role of a broker is to act as a 'trusted intermediary' who can provide some form of expertise to help with choosing a service that we need. The idea of 'insurance broker' and 'mortgage broker' remain relevant, even in an era of internet shopping (and many web sites claim to offer the services of a broker). We turn to a broker so that we can draw upon their expert domain knowledge to determine which provider is likely to be best suited to our needs.

The same can occur in a software context, particularly in the context of service-oriented architectures. In this context we have *clients* (which themselves may be providing services to others) and *servers* and these need to be matched so that clients obtain the most appropriate service for any request (Bennett, Layzell, Budgen, Brereton, Macaulay & Munro 2000).

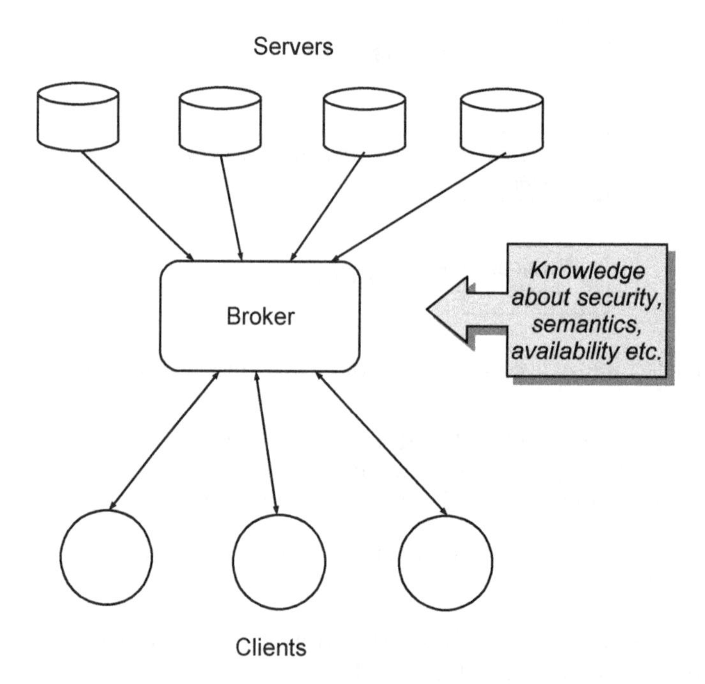

FIGURE 15.6: A simple broker architecture

When using a *broker* architecture for software, the client sends requests to the broker, the broker seeks out and uses the most appropriate service, and then returns the outcomes to the client. This involves a *dynamic binding* for each request (or set of requests), which is managed by the broker so that the client may well not be aware of which server provided the necessary response. Figure 15.6 shows this schematically. The broker architecture is commonly associated with platforms for distributed service provision such as .NET and Enterprise Java Beans (EJB). This is because the broker pattern is particularly useful in a dynamic environment, where the set of service providers (which may include 'information services') and their availability may be subject to regular change, as may well occur with distributed systems.

Use of a broker does add a run-time overhead in exchange for this decoupling, and incorporating security needs can add further to this (Turner, Brereton & Budgen 2006). The broker itself needs to be secure (and hence trustworthy) and also any requests and responses must be transmitted securely. Equally, a broker does provide a robust run-time environment since the broker is able to make substitutions for a server that has failed, or that may have become available.

15.3 Design patterns

Design patterns address smaller and more detailed pieces of an overall design than architectural patterns and have been adopted particularly widely in the object-oriented context. As we noted when looking at plan-driven approaches to design, this form of 'design method' tends to become overly-complex when used with objects, and design patterns, rightly or wrongly, have been seen as one way of overcoming the problem by reusing 'proven' solutions to specific aspects of an ISP.

Objects offer a useful form of design element that can form one of the components of a pattern. As we noted in Chapter 10, an object:

- possesses a *state*, which is encapsulated within the object and can be inspected and changed through its methods;
- exhibits *behaviour* through its responses to external events;
- possesses an *identity*, since more than one object may be created from a class.

Objects are also capable of using static and dynamic binding (valuable for patterns) and can be coupled to other objects using a number of different mechanisms. One thing that we might note though, as already mentioned, is that design patterns tend to make use of aggregations of objects rather than employing inheritance. Employing inheritance remains a design challenge, much as it did for plan-driven design methods, and opportunities to employ inheritance may still need to be identified through techniques such as refactoring.

For our purposes, the important features of an object are that it provides an *abstraction* of some aspect of the design model, that it incorporates *encapsulation* of state information, and that it provides externally accessible *methods*.

While design patterns provide some 'part-solutions', a really major reason for using these, and one that is quite different from the purpose of architectural patterns, is that they are intended to be used to address those aspects of a design that are most likely to undergo *change*. Much of the rationale behind different patterns is to isolate such elements in the design model with the goal of reducing the impact of future evolution. This is something that we will particularly examine when we look at some examples of patterns. It also means that patterns often seek to employ loose coupling to provide the necessary flexibility.

A text that helped popularise the idea of the design pattern is the pioneering book from the 'Gang of Four' (often abbreviated to *GoF* in the patterns literature): Erich Gamma, Richard Helm, Ralph Johnson and John Vlissides (1995). Their book provides a catalogue of 23 design patterns and helped establish ideas about pattern classification and description. Valuable though this is, it is not really a tutorial about patterns, and a text such as *Head First Design Patterns* by Bates, Sierra, Freeman & Robson (2009) provides a much easier (if sometimes rather eccentric) way of learning about the design pattern concept and about some major patterns.

Design patterns have attracted a strong and involved user community, and a major resource providing various forms of knowledge about design patterns is provided by the web site at www. hillside.net. This on-line resource provides details of many patterns, as well as links to other books and tutorials.

The adoption of a pattern within the design of an application does represent an important long-term decision (which of course, implies an element of technical debt). Where an application evolves through time, one of the risks is that the changes to the objects mean that a pattern may decay, or that its presence might even impede unforeseen forms of change. There isn't much that can be done about the latter (or they wouldn't be unforeseen), but the issue of decay is one that designers need to consider when making changes.

Two useful concepts related to pattern decay are those of *grime* and *rot*, that are used to describe the corruption of a pattern that can occur when changes are made to the objects involved in the pattern (Izurieta & Bieman 2007, Feitosa, Avgeriou, Ampatzoglou & Nakagawa 2017). *Pattern grime* is the "degradation of design pattern instance due to buildup of unrelated artifacts in pattern instances" and in essence involves material being added to the classes and objects of a pattern that are not directly related to the purpose of the pattern. *Pattern rot*, which is less common, is the "deterioration of the structural or functional integrity" of the elements making up a pattern. The presence of grime and rot may mean that a pattern cannot support changes to the application in the way that was originally envisaged when the decision was made to adopt that pattern.

Patterns can take a variety of forms, and the *GoF* classified patterns along two 'axes', as illustrated in Figure 15.7, which were as follows.

1. The *purpose* of the pattern. Its purpose describes what a pattern is used for, and is usually described as being one of the following three types.

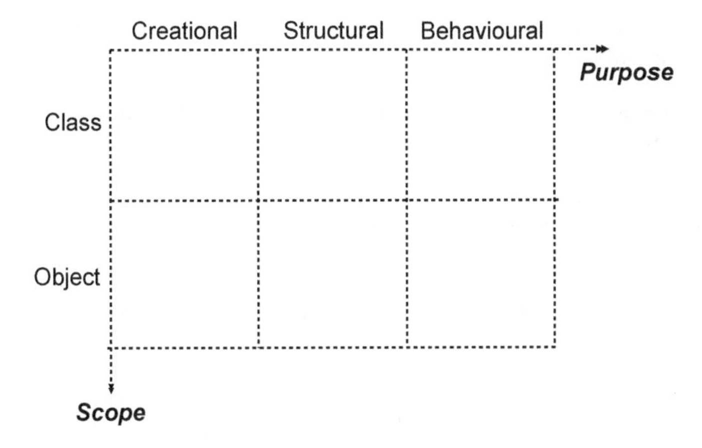

FIGURE 15.7: Pattern classification scheme used by the 'Gang of Four'

- *creational* patterns are concerned with occasions where objects need to be created in order to achieve the purpose of the pattern;
- *structural* patterns address the way in which the constituent classes or objects are composed;
- *behavioural* patterns describe the ways that classes or objects interact, and how responsibilities for different aspects of the pattern's purpose are allocated between them.

2. The *scope* of the pattern. This describes whether the pattern is primarily one that addresses the use of *classes* or the use of *objects* (where we can regard a class as being a form of 'template' from which one or more instantiating objects are realised). Most patterns deal with objects, and so we concentrate on looking at examples of these.

In terms of the viewpoints model, creational and structural patterns are essentially ones for which the organisation of the pattern is associated with the *constructional* viewpoint. And obviously, the organisation of behavioural patterns is associated with the *behavioural* viewpoint.

While we use the framework shown in Figure 15.7 here, it is worth noting that others have employed other ways of categorising patterns. Although in Buschmann et al. (1996) the main focus is on architectural patterns, there is also some discussion of a number of design patterns. They categorise these using a set of role-based headings (structural decomposition, organisation of work, access control, management and communication).

The remaining question to address, before looking at some examples of patterns, is how to describe a pattern. The books by Gamma et al. (1995) and Buschmann et al. (1996) use slightly different templates for this. The following template is something of a merging of these (and others) to try and capture all of the issues that need to be documented, and one that we use in the following subsections.

Name. This is used to identify a pattern and (ideally) to indicate the essence of what it does.

Also known as. Rather pragmatically most catalogues of patterns do recognise that they may be categorised using different identifiers in other contexts.

Problem. This is the design 'problem' that the pattern is intended to address.

Solution. This outlines the way in which the pattern addresses the problem, and explains any design principles that underpin the solution.

Example. Provides a real-world example of the problem and of the pattern-based solution.

Applicability. Describes the situation in which the problem might arise and where it may be appropriate to employ the pattern, possibly including hints for recognising such situations.

Structure. Provides a detailed description of the organisation of the solution usually employing diagrammatic forms to model both the *behavioural* and the *constructional* viewpoints.

Implementation. A set of guidelines for implementing the patterns, noting any possible pitfalls that might arise.

Known uses. A set of examples from existing systems.

Related Patterns. (Also known as **See Also**.) Patterns that address similar problems or that complement this one in some way.

Consequences. Design trade-offs that might need to be made when employing the pattern, as well as any constraints its use might impose and possible forms of technical debt that it might incur.

The choice of patterns described in the following subsections is based upon those *GoF* patterns that were identified as being most useful in the survey of pattern experts described by Zhang & Budgen (2013). (This survey is discussed more fully in the section on empirical evidence.) And as a final comment, when discussing *GoF* patterns it is common practice to identify the page number where the description of the pattern begins, and this practice has been followed here.

15.3.1 Proxy (207)

The role of a *proxy* can be undertaken in various ways as part of everyday life, when someone acts 'on behalf of' another person for some purpose. A common example is in voting, where someone unable to go and take part in a ballot will appoint another person to act as their proxy and cast their vote for them. This pattern employs the same concept, and involves one object acting on behalf of another object.

The *GoF* classify this pattern as being *Object Structural* whereas Buschmann et al. describe it as being *Access Control*. Using the template outlined above this is described as follows.

Name. Proxy

Also known as. Surrogate

Problem. Proxy addresses the problem where direct provision of access to an actual object requires to be moderated in some way. This may be because access represents a significant overhead of some form (time, space) or because different types of client may require or be entitled to different levels of access.

Solution. This is to provide a *representative* of the object and let the client object communicate with this proxy rather than with the actual object. The proxy provides the same interface as the actual object and ensures correct access to it, while possibly performing additional tasks such as enforcing access protection rules.

Example. The *GoF* provides the example of an object representing a complex image when used in a word processor. Rather than incurring the overhead of loading a (potentially) large graphical object it may be possible to employ a much simpler proxy that ensures that key properties (such as position and boundaries) are correctly represented in the screen image. With more widespread access to electronic data sources, a more common example today is probably the *protection proxy* form, in which the proxy is used to control the level of access that users may be permitted to have to an object or to information. For example, in the case of access to Electronic Health Records (EHR), a doctor is likely to have fuller access than (say) a nurse, and an administrator might have quite limited access that is restricted solely to demographic information.

Applicability. Proxy can also be employed to control access to relatively complex objects, which may be in a different space (remote proxy); with minimising overheads, as in the case of the word processor and image (virtual proxy); and controlling different access rights (protection proxy).

Structure. Figure 15.8 illustrates the idea of proxy using a class diagram. Normally, requests from the client will be serviced by the methods of the proxy, relaying these on to the original object only when necessary. Figure 15.9 uses a sequence diagram to illustrate the operation of virtual proxy (or alternatively this could be viewed as an example of protection proxy, with the first request being refused for lack of access permission).

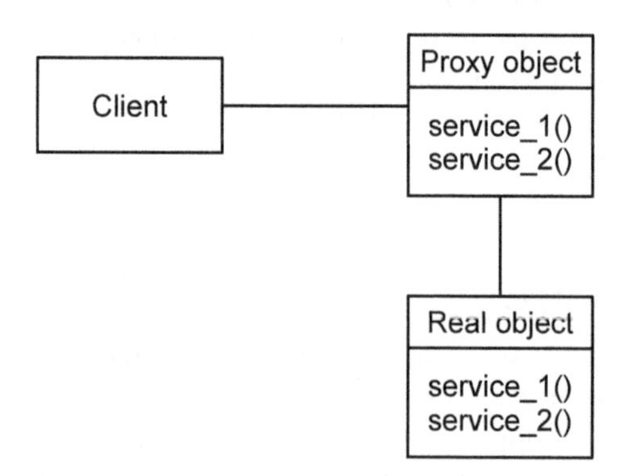

FIGURE 15.8: Class diagram for the proxy design pattern

Implementation. Since this is not relevant to the theme of the book, we omit this element.

Known uses. Both the *GoF* and Buschmann et al. provide examples, with the latter being more extensive and varied and also relating more strongly to internet-based forms of implementation, where this pattern is quite widely employed.

Related patterns. The *Adapter* (139) and *Decorator* (175) patterns are also concerned with issues related to interfaces.

Consequences. Obviously, the use of proxy creates an overhead by adding a level of indirection when accessing potentially quite important objects.

As an initial example of a design pattern, *proxy* illustrates a number of the points made above about patterns in general. It is compositional in form, addressing what is very much a sub-problem for any design, relatively simple in its structure, and provides quite loose coupling that may help with future changes. It is also a pattern where ideas from 'conventional' coding and the internet in its various forms coincide, showing how the concept behind a pattern can apply across quite a wide range of roles and forms.

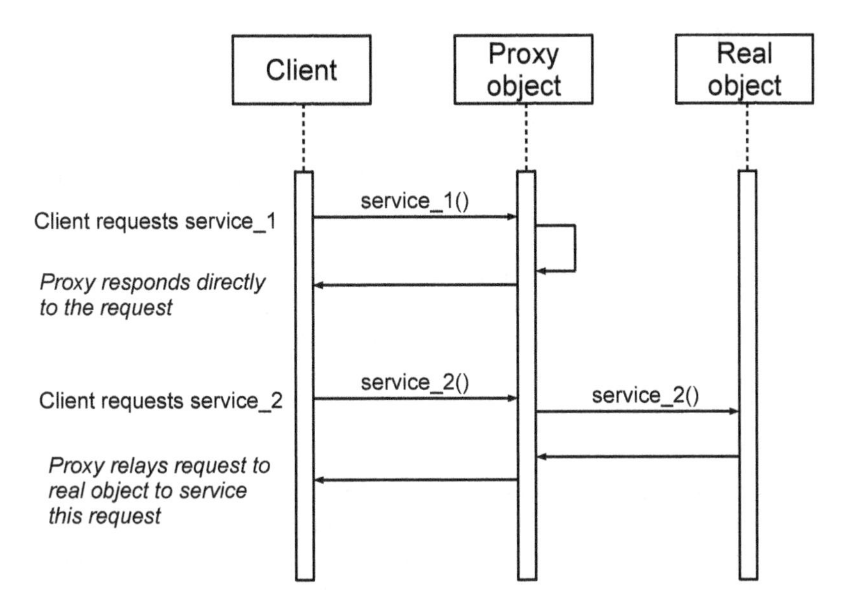

FIGURE 15.9: Sequence diagram for the proxy design pattern

The proxy pattern in the CCC

An obvious role for the *proxy* pattern within the CCC is to provide a protection proxy role for access to car objects. There are a number of roles in the overall CCC organisation that need access to car objects, but for different purposes. Customers need access for bookings, the maintenance team need access to monitor the state of a car and to be able to withdraw it from use if necessary, the billing system may need further access to details of journeys and so on.

While it would be possible to build this into the car object itself, it would then mean that the car object would require knowledge about such roles, which in themselves, have no direct relevance to its function. So, following the principle of *separation of concerns*, it is better to use a proxy as the means of embodying the knowledge about who may access what. It also aids future evolution of the application, which might create new roles, or roles that are specialisations of existing roles. Such developments would then only affect the proxy object.

15.3.2 Observer (293)

Observer is a well-known and widely-used pattern and is categorised as being *object behavioural* by the *GoF*. It embodies the concept of a *publish-subscribe* relationship between the design elements. The pattern as described by the *GoF* is fairly basic, omitting various 'housekeeping' issues, and the description provided here likewise concentrates on the core model, while indicating where additional aspects need to be considered for implementation.

Name. Observer

Also known as. Dependents, Publish-Subscribe

Problem. A common side-effect of organising a software system as a set of co-operating objects is the need to maintain consistency between related objects, while avoiding overly-tight coupling that may constrain performance as well as reduce the scope for reuse.

Solution. This involves creating a 'publish-subscribe' relationship between the *subject* and the *observers*. Whenever the subject undergoes a change of state it notifies the observers that this has occurred, and they can then query the subject in order to synchronise with it and obtain any updated values.

Example. A simple example is that of a spreadsheet. When the data in a cell or group of cells is changed, any graphs, pie-charts or the like that are using that data will also need to be redrawn. Here the spreadsheet cell is the subject, and the objects that are responsible for drawing the charts and graphs are the observers.

Applicability. This form of relationship occurs widely wherever objects are working together to perform some task, and where an object needs to be able to notify other objects about changes without any knowledge about those objects.

Structure. Figure 15.10 illustrates the idea of observer using a class diagram. An interface is used to implement the subject-observer relationship and this is then implemented by the concrete subject and observers. Observers register with the subject to receive calls to `update()` from the subject. On receiving an `update()` message it is then the task of the observer to use some form of `getState()` to check if the change affects it (since observers may be interested in different aspects of the state, they may use different versions of `getState()`). Figure 15.11 uses a sequence diagram to illustrate the operation of observer.

Implementation. Again, this element is omitted here.

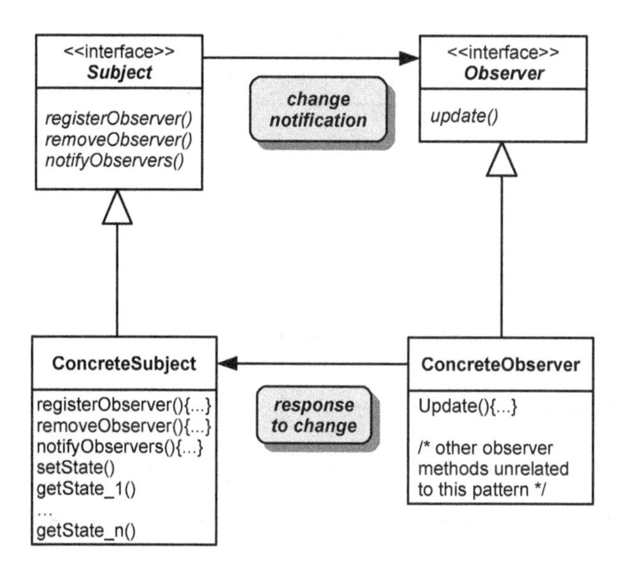

FIGURE 15.10: Class diagram for the observer design pattern

Known uses. *Observer* is a very widely used pattern. The example above of the spreadsheet is just one of many. Another well-known one is within the MVC architectural pattern, where the model acts as a subject, and the views act as observers, being notified when changes are made to the model.

Related patterns. *Publish-Subscribe*, *Mediator*.

Consequences. While *Observer* provides loose coupling that makes it possible to vary subjects and observers independently, and to add observers without needing to modify the subject or other observers, there are some 'housekeeping' issues that need to be considered. These include:

- a subject does need to keep track of the observers;
- if an observer is observing more than one subject then it is necessary to extend the `update()` method so that the observer is able to know which subject has provided the notification;
- if a subject is deleted then it is necessary to notify all of its observers to avoid dangling references.

The model shown in Figure 15.10 describes the very basic operation (largely ignoring the housekeeping issues above). Even so, this is a relatively simple as well as effective pattern (it was the pattern that was most highly valued in the survey by Zhang & Budgen (2013)).

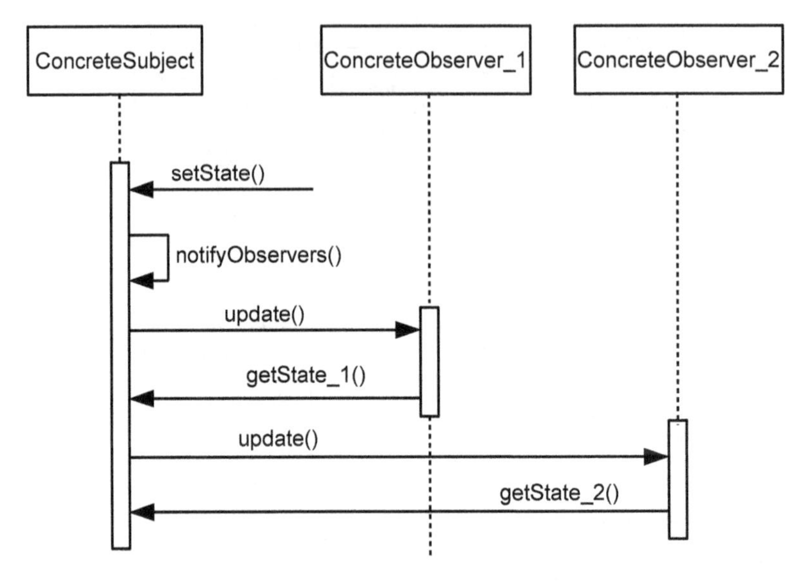

FIGURE 15.11: Sequence diagram for the observer design pattern

The Observer pattern in the CCC

One of the most obvious roles for the *Observer* pattern is to help monitor the status and use of car objects. When a car object changes its status (such as becoming *available*), then it can 'publish' that change of state, and any subscribers can take note. Different subscribers likely to be interested in car states will include club members seeking a nearby car, the maintenance team, and the billing system.
When seeking a car, a club member may become an observer for several cars. But once the member has selected one, then the appropriate booking object needs to remove itself from the list of subscribers to any other cars in order to avoid 'dangling references'.

15.3.3 Abstract Factory(87)

The *Abstract Factory* provides an example of an *object creational* pattern that addresses a problem that probably is a bit more limited in scope than those of the previous two examples, although the problem itself is an important one. The aim of this pattern is to make a 'client' class independent of the specific set of objects that may need to be created for it to perform its task (decoupling it from them), where there is a choice of classes that can be instantiated. The decision about which class to instantiate is deferred to runtime, when a specific concrete factory class is used to create the objects.

Name. Abstract Factory

Also known as. Kit

Problem. This pattern addresses the need to achieve portability of an application across environments. A common role is to enable an application (the 'client') to be used on a range of platforms that provide different forms of 'look and feel' user interaction.

Solution. The abstract factory defines an interface for creating each basic form of widget used in interaction (such as scroll bars, radio buttons etc.). Clients perform operations to obtain a new instance of a widget, but are not aware which concrete class is being used.

Example. The widgets will usually form a user interaction toolkit that supports multiple standards for such widgets as scroll bars, buttons etc.

Applicability. The roles for this pattern address situations where either:

- a system needs to be independent of the way in which its products are created, composed and represented; or
- a system may be configured locally, using just one form from a family of products.

Structure. Figure 15.12 illustrates the idea of abstract factory using a class diagram. The client uses the *abstract factory* when it is compiled and then at run-time it is linked to an actual (concrete) factory. So if *ConcreteFactory1* is selected, that class will provide *ProductA* and *ProductB* for use by the client. (Of course, the interfaces that these must conform to are specified by the abstract factory.)

Implementation. Again, we omit this element.

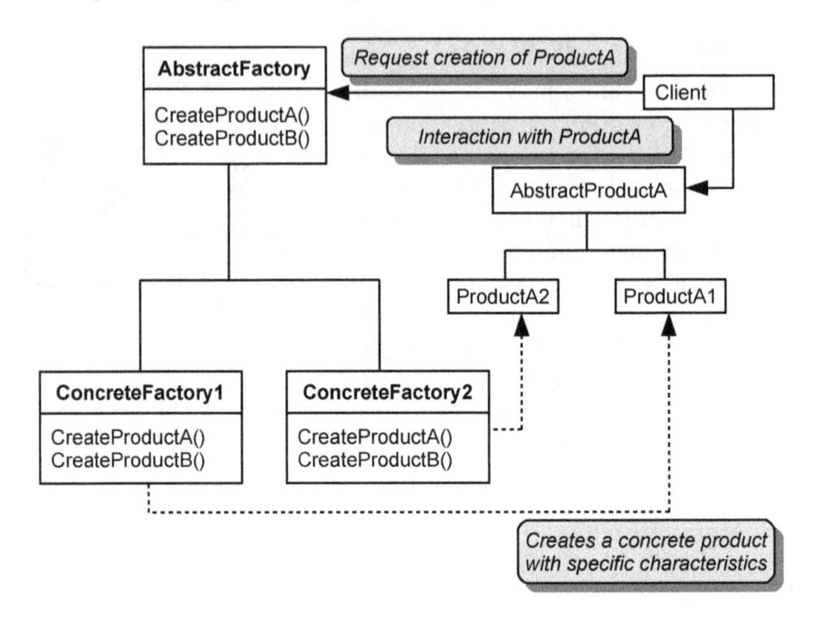

FIGURE 15.12: Class diagram for the abstract factory design pattern

Known uses. *Abstract factory* isolates clients from the responsibility and process of creating product objects, the client manipulates the instances through their abstract interfaces and hence remains unaware of the particular implementation.

Related patterns. Prototype

Consequences. While the abstract factory can be easily extended to include new members of a product family (for example, a new form of user interface), adding new products to the set provided by the factory is not easy, because the pattern defines a fixed set of products. So adding (say) a variant form of radio buttons would require modifying the abstract factory and recreating all of the concrete factories. In that sense, using abstract factory does create a form of technical debt through the constraints this creates.

Abstract factory meets a quite specific need and is generally considered to do it well. However, apart from the example of different user interfaces it is not as likely to be so widely used as the previous two examples.

15.4 Other uses of patterns

This section addresses two rather different pattern-related issues. The first is to briefly examine an example of how design patterns can be used with another architectural style (SOA), while the second involves looking at how patterns can be employed to describe experiences of what *doesn't* work.

15.4.1 Software service patterns

Software service concepts were described in Chapter 11. While components (which can be considered as the precursors of services) do not seem to have attracted much interest from the patterns community, software services have presented a topic of greater interest, and indeed, patterns may well be a particularly effective way of adapting the service concept to user needs.

At the architectural pattern level, Bass et al. (2013) consider that the *service-oriented architectural pattern* describes the use by service consumers of distributed services offered by service providers. Essentially this pattern describes the use of services in the form described in Chapter 11. It identifies the need for a service to publish details of what is provided, and also to address quality issues through the use of *service-level agreements* (SLAs), which reflects the strong business element often associated with the use of the service model.

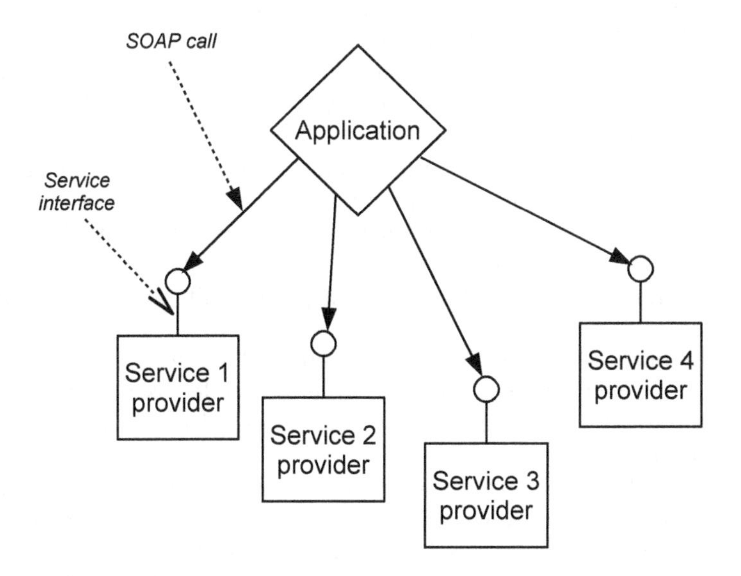

FIGURE 15.13: A very simple SOA pattern

Topologically, this architectural pattern can be expected to possess a 'star' configuration. A simple example is provided in Figure 15.13. In this example the external service providers are essentially 'fixed' and some form of 'service broker' or 'orchestration server' could well be included to allow greater flexibility of the choice of providers and configuration.

At the more detailed level (akin to the *GoF* patterns), a number of books have documented a wide range of patterns. A major source is the text by Erl (2009). While some patterns are derived from the object-oriented pattern set, most appear to be unique to the service context. Service patterns are complicated by a number of factors, including the existence of different service frameworks, the dynamic (self-adapting) nature of many service applications, and the lack of standard notations.

15.4.2 Design anti-patterns and code smells

Ideas about software reuse, which is a key motivation for employing design patterns, has understandably focused upon finding ways to reuse *good* experiences. But of course, a designer's knowledge schema may well include experiences that were unsuccessful too, and these also need to be remembered so that they can be avoided in the future. So, not surprisingly, although less well-codified than the concept of the design pattern, this has led to the concept of the *design anti-pattern.*

In a very readable, if rather tongue-in-cheek, paper, John Long (2001) provides some examples of "obvious, but wrong, solutions to recurring problems", putting particular emphasis upon the processes involved in choosing to use them. Indeed, the anti-patterns literature does tend to put more emphasis upon the reasons *why* wrong solutions are adopted, rather than on the specific form of the wrong solutions themselves (Brown, Malveau, McCormick & Mowbray 1998).

This emphasis is quite understandable, since technical solutions are rarely completely wrong in themselves, although a particular solution may well be the wrong one to adopt in a particular context. So, while the literature on design patterns emphasises *solution structures*, although recognising the influence of context and motivation, the anti-patterns literature focuses mainly on *motivation.*

For our purposes, the main message here is that *reuse* of a design model (or in the case of patterns, of part of a design model) is not automatically "a good thing". What has proved successful in one situation may not always work in another.

However, while anti-patterns are in some ways something of an evolutionary dead end, they can be considered as a form of antecedent for the much more widely adopted idea of *code smells.* The term 'code smell' is generally attributed to Kent Beck, and in the absence of a formal definition (the concept does have a subjective element arising from context), the description provided in Fowler (1999) is quite widely used. In this, a code smell is considered as being a structural weakness that indicates a deeper flaw, and that needs to be removed from the source code through the use of *refactoring* in order to improve the maintainability of the software. (While the term does provide an unquestionably evocative description of the sort of issue involved, it is probably better not used when explaining refactoring plans to project managers, customers etc.)

As the term indicates, code smells are likely to be recognised during implementation, although some may well be detectable in advance of that. Here we briefly look at some examples.

God Class. A *God Class* "refers to those classes that tend to centralise the intelligence of the system. An instance of a god class performs most of the work, delegating only minor details to a set of trivial classes and using the data from other classes" (Lanza & Marinescu 2006). A God Class may well make frequent access to the data of foreign classes and its role violates the object-oriented design principle that each class should only have one responsibility. God Classes also tend to be very large, affecting ease of comprehension. Because of its role, it is anticipated that such a class will be changed frequently during maintenance, and hence more likely to include errors.

However, the study by Olbrich, Cruzes & Sjøberg (2010), looking at the evolution of three OSS applications does demonstrate that it may well be that the use of a God Class is a quite reasonable way to organise some forms of application, provided that its size is not extreme. This emphasises the point made above that the identification of relevant code smells may well be subjective and contextual.

Data Class. Such a class has data fields, together with methods for changing and inspecting these, but plays no role in the operational purposes of the application. Its lack of a behavioural role weakens the object-oriented structures and at the worst it may simply be a mechanism for including global data. As observed in (Fowler 1999) this may be an acceptable form for a class to have during initial development, but at some point it should take on the role of a 'grownup'.

Code Duplication. Duplication of code can arise in more or less any architectural style (for example, it can easily arise in the lower levels of a call-and-return structure). Duplication can lead to problems with maintenance, particularly where only one instance is changed, creating inconsistencies. Fowler (1999) suggests some techniques to employ for refactoring when this is detected.

Looking at these examples, although they may (partly at least) arise from design decisions, their characteristics relate strongly to code. Hence it may be more difficult to identify these in a design model (the *God Class* perhaps excepted).

Where code smells do have a useful role is when identifying the need for possible *refactoring*. As discussed in Section 14.7, refactoring involves reorganisation of the code from a constructional perspective, while retaining its function and behaviour, in order to help with the future evolution of an application. Code smells can help with prioritising the need for such change in the design. They may also provide a motivation for refactoring that is related to the design model itself, although they still have a strong link to the actual code structures.

15.5 Designing with patterns

Having reviewed various examples of the pattern concept, the next obvious question to ask is "how do we use design patterns to solve design problems?". In this section we discuss some issues related to pattern use.

Many of the books that describe design patterns, such as those by Gamma et al. (1995) and Buschmann et al. (1996) are very much structured as catalogues of patterns. And later texts such as Bates et al. (2009) focus largely on the issues of how to implement patterns. Important as both of these are, particularly for documenting patterns, they still leave our initial question unanswered. Unfortunately, rather as a gardening catalogue, full of glorious colour pictures of healthy, thriving plants, provides little real aid with the task of planning a new garden—although it may tell us which plants like shade and how tall they will grow—so it is apt to be with design patterns. Possession of a catalogue provides a source of ideas; it provides information that can help with planning and anticipating possible consequences, but the task of working out how to use its contents is still a creative activity.

(Actually, the analogy with planning a garden is quite a good one, at least, if you like gardens, since gardens do evolve and exhibit *behaviour*, even if over much longer periods of time than we expect with software. Trees grow and shade different parts of the garden, some plants take time to become established but then take over adjacent sections unless controlled. So, like the software designer, the gardener's planning requires an ability to envisage some future state, while having inadequate control over both the conditions and the quality of materials that they need to use to achieve that state.)

How then do we use catalogues of design patterns, whether in books or on-line? Well, the *GoF* advice is very much along the lines that patterns need to be *learned*, and that by studying patterns the designer will acquire both insight into how and why the pattern works as well as enough familiarity with it to be able to recognise those situations where it can be used to effect. Designers are also advised to follow the following two principles.

- *Program to an interface, not an implementation.* What this means is that a client object should not be aware of the identity of the actual object that is used to service a particular request, so long as its form and behaviour conform to the interface specification.

- *Favour object composition over class inheritance.* This is not to deny the value of inheritance as a mechanism for reuse, but rather that it should not be over-used.

This implies that the minimum conditions for the successful use of patterns require that a designer should:

- acquire a 'vocabulary' of design patterns;
- be able to recognise where a particular pattern could provide a useful solution;
- have an understanding of how to realise the pattern within that context.

Given that there are now hundreds of documented patterns (the book by Gamma et al. (1995) documented only 23 of them), these are quite challenging requirements, especially given that the question of which patterns are the ones that are most valuable to learn (first) is not easily answered.

The strategy advocated by Buschmann et al. (1996) is rather different, although the basic conditions for when it is appropriate to employ a pattern are the same. They advise classifying a given problem in the same way that the patterns themselves are classified, as a step towards being able to identify potentially useful patterns. Their basic process is as follows.

1. Specify the problem, and if possible, any sub-problems involved.
2. Determine the category of pattern that is appropriate to the design activity being performed (that is, *architectural* or *design* patterns).
3. Determine the problem category appropriate to the problem.
4. Compare the problem description with the available set of patterns in the catalogue that fit that problem description.
5. Consider the benefits and liabilities (assess the design trade-offs and possible technical debt that will be incurred by using the pattern). Of course, this may also need to include the possibility that there is no existing tried and tested pattern for the given problem.
6. Select the pattern variant that best fits the problem and offers most benefits.

Since their strategy includes architectural styles as well as design patterns and idioms, the above process can be considered as fairly comprehensive, and in principle at least, avoids the need to learn a growing catalogue of patterns. We might also note that it has a distinctively top-down decompositional aspect, which is perhaps unexpected, given that patterns are essentially a compositional concept. Perhaps the main practical limitation is that pattern documentation has tended to follow the structure adopted by the *GoF*.

If we go back to the model of pattern use represented by Figure 15.2, an obvious question is how to recognise the opportunity to use a pattern. We should not assume that there is always going to be a suitable pattern when using a process such as the one above. Indeed, trying to force all problems into a 'pattern framework' is likely to produce exactly the opposite of what it intended. So step 5 in the above process is an important one, and the designer may well need to be able to recognise when there is no 'ready-made' pattern that fits a given problem. A key thing here is recognising that a particular design problem is likely to have been encountered in enough situations to be likely to have resulted in someone creating a pattern for it. Again, this is where the more experienced designer is likely to have an advantage.

The process recommended by the *GoF* carries within it a self-limiting aspect that is clearly difficult to overcome. This arises from the way that, as more patterns are identified and added to the corpus of pattern knowledge, so the ability of the individual designer to *learn* all of these becomes an ever-increasing challenge. What is needed is some agreed way of indexing patterns, or at least agreement about what constitutes a core set of patterns that everyone might be expected to learn. Neither of these has really emerged, and in their absence, the process recommended by Buschmann et al. would seem to cope better with the nature of ISPs.

15.6 Empirical knowledge about designing with patterns

While the design patterns community has contributed some very valuable insight into how we can reuse design ideas, this has not always been quite as balanced as might be desired. Enthusiasts continue to write patterns and encourage others to do the same; there are conferences on patterns (usually labelled as *xxx*PLoP, where PLoP stands for 'Pattern Languages of Programs'); and a journal (TPLoP of course). What has been less evident is any enthusiasm for winnowing out the less useful patterns and evaluating the concept in general. Patterns are unquestionably a useful concept, but as with all concepts employed for design, it is important to obtain an understanding of their limitations too.

This section almost entirely describes studies related to object-oriented design patterns. There is very little empirical research into architectural patterns (perhaps not entirely surprising) or into the use of design patterns with other architectural styles.

Zhang & Budgen (2012) conducted a systematic review into what was known about design patterns. They found a small number of empirical studies (mainly small-scale experiments) and a number of 'experience papers', with that by Wendorff (2001) being particularly insightful. All of the patterns studied were either directly taken from the *GoF* or were close derivatives. They concluded that: "we could not identify firm support for any of the claims made for patterns in general, although there was some support for the usefulness of patterns in providing a framework for maintenance, and some qualitative indication that they do not help novices learn about design". However, the strength of evidence for the overall findings from the study was limited by the small number of primary studies and the way that these used a spread of different patterns, with *Composite*, *Observer* and *Visitor* being the only patterns that had been addressed by more than two or three primary studies.

This study was followed up with two surveys. The first was a quantitative survey of the usefulness of the 23 *GoF* patterns, using the authors of pattern papers as the sampling frame (Zhang & Budgen 2013), and then was augmented by a more qualitative survey (using the same sampling frame), seeking explanations for some of the results (Zhang, Budgen & Drummond 2012). As with almost any survey undertaken in software engineering, it is extremely difficult to determine to what degree the sampling frame used could be considered as being representative of pattern users as a whole.

From the first survey, the most highly rated patterns were *Observer*, *Composite* and *Abstract Factory*, and there was also a substantial group that were not considered to be useful (most notably, *Memento*). There was also considerable ambivalence about *Singleton*, which was investigated further in the second survey. The *Singleton* pattern is a creational pattern that "ensures a class only has one instance", with a single global access point. (A good example of a situation where the use of this might be appropriate is for a spooler class responsible for managing output to a printer.) The qualitative survey revealed that a significant number of users were concerned that *Singleton* was easily mis-used to provide global variables, and that its use also increased coupling. So, while it could be useful for limited roles, the potential disadvantages in terms of introducing undesirable features into a design model were such that they felt it better to avoid its use[1].

While there have been quite comprehensive studies of research trends related to the use of design patterns, such as Mayvan, Rasoolzadegan & Yazdi (2017), (which found that pattern development, pattern mining, and pattern usage were the most active research topics in the field of design patterns), there appear to have been relatively few studies related to any form of evaluation of the patterns themselves. Given how much patterns are valued by quite a wide community (and lack of research does not imply lack of value), this lack of critical analysis is disappointing.

This situation is even more marked for the use of software service patterns. Although there is no lack of tutorial material, there seem to have been few studies that assessed the usefulness of this class of patterns.

[1]For these reasons, this has not been used as one of our examples of patterns.

If we turn briefly to the anti-pattern context, then we find that, despite the issues of context and difficulty of definition, there have been many studies on the use of *code smells*. An example of such a study is that by Palomba, Bavota, Penta, Fasano, Oliveto & Lucia (2018), and it is worth noting that this, like many others in this area, largely draws upon data from open source projects. However, these studies do not appear to offer any very conclusive views on the usefulness of code smells, and rarely provide any systematic form of evaluation. Some of the cause of this may be the influence of *context* as a confounding factor, as noted by Olbrich et al. (2010).

Key take-home points about designing with patterns

Design patterns provide a useful and valued means of codifying and exchanging information about design structures (large and small) that have been found to be effective by software designers.

A pattern describes both a 'design problem' and a generic solution to that problem that can be reused over and over again in many different ways. Labelling of patterns aids with knowledge transfer and reuse.

Patterns are more concerned with coupling through composition than with inheritance. They generally aim to provide loose coupling between design elements so as to aid the 'evolution' of a design model.

Architectural patterns describe ways of structuring applications to fit particular characteristics of their role.

Design patterns provide 'part-solutions' for elements of a design model, simplifying the process of 'solving' an ISP by providing a way of organising specific aspects of the design model.

Patterns are used for different aspects of the design model. *Creational* and *structural* patterns are organised around the constructional viewpoint, while *behavioural* patterns structure the design within the behavioural viewpoint.

Patterns can be used with any architectural style (in principle) but have largely evolved for the object-oriented architectural style where there is a large community of pattern enthusiasts.

The design process when making use of design patterns is an informal one with only quite limited guidance about the recognition and use of patterns being available to help the designer.

Empirical knowledge about the usefulness of design patterns and architectural patterns is limited. However, it is clear that some patterns (most notably *Observer*, *Composite* and *Abstract Factory*) are valued by experienced designers. There is some evidence that the use of patterns is unlikely to provide help for less experienced designers.

Chapter 16

Designing with Components and Services

In Chapter 11 we looked at component technology and the concept of the software service. Our concern there was to describe their forms and to consider the properties of components and services that were of importance from the perspective of how they might be modelled.

This chapter looks at how components and services can be *composed* to create applications, and some factors that we might need to consider when designing with such elements.

16.1 Modular design

Chapter 11 addressed the question of how to design components, both in the form of 'local' software components, and also software services (which of course, can be viewed as being a form of 'remote' component). In this chapter we briefly discuss the question of producing designs for building applications that make use of both components and services.

The two forms have much in common as well as some differences when it comes to producing a design. They particularly share a 'plug-and-play' philosophy, arising from the widespread use of dynamic binding when *composing* an application from existing elements, hence the reference above to the idea of 'modular' design. And despite the importance of *reuse* in software development as a whole, in neither case are there any well-established practices to assist with formulating an overall design model around the concept of reuse. Indeed, up to this point, the ideas about creating software applications presented in Part III have essentially assumed a 'bespoke' development process, with the elements making up each application (or product line) largely being different and unique. So, while the development of ideas about component-based software engineering (CBSE) could build upon experiences with using objects, those experiences provided little opportunity for accumulating and formalising declarative and procedural knowledge about modular design.

However, when creating applications from existing components and services, some aspects of *incremental design* may well be adapted for this purpose, because both forms provide scope for the use of *prototyping*, particularly *exploratory* and *evolutionary* prototyping. Exploratory prototyping allows the designer to investigate the use of different component combinations while developing an application, while evolutionary prototyping may provide an overarching development context in which the application emerges from a series of intermediate forms.

The use of prototyping rather than the development of more structured design models may however create limited opportunity for providing knowledge transfer that has been derived from experience of composing components. This is because, although components and services are implicitly well-suited to the use of a *compositional* strategy, the major influence upon design decisions is likely to be the identification of what modules are available from the *component catalogue* rather than design strategy.

16.2 Designing with components

As noted above, there is very little guidance available about how to develop a design by using components. This is true even for a well-established architectural form such as JavaBeans, the component architecture used with the Java platform. The component composition process itself is further complicated by the way that components can be used to perform what we might term *horizontal* roles within a system (whereby the overall functionality of the application is distributed between them), and also *vertical* roles in which they provide 'layers' of services through component frameworks such as CORBA and .NET. The latter role essentially corresponds to the use of a *Layers* architectural pattern. This is illustrated in Figure 16.1.

Our concern here is primarily about how the design choices related to 'horizontal' structuring are to be made, on the basis that these provide the detailed design, whereas the 'vertical' structuring provides an architectural design. But of course, as always, the two cannot be regarded as entirely separate.

Where the aim is to develop an application by reusing existing components, then in selecting a set of components that can be composed to make up the overall functionality of a system, there are two distinct strategies that might be employed. (These can be viewed as forming extremes of course, and composition may well be undertaken using a mix of these.)

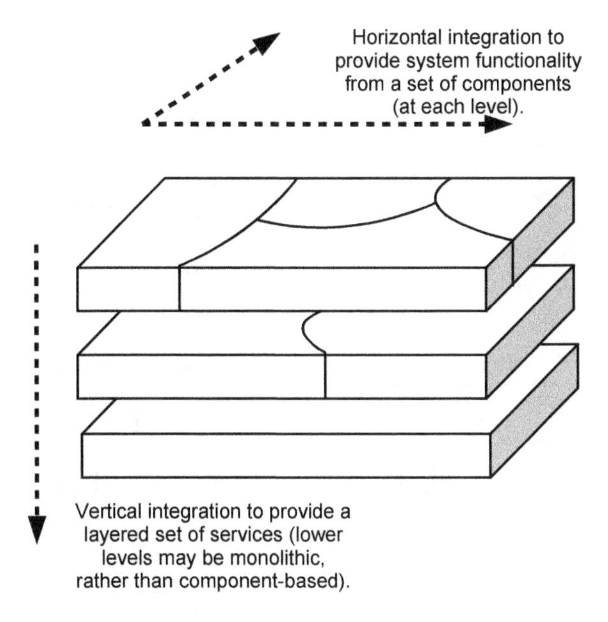

FIGURE 16.1: Horizontal and vertical integration of components

- Identify the general needs of the application problem, and search for a set of components that collectively match that functionality, and seek to construct the design (and implicitly, the application) by aggregating these in some way. (This can be termed as an *element first* strategy.)
- Decompose the application problem into well-defined sub-problems that address specific tasks and then seek a set of components that will fit the needs of the individual sub-problems. (This can be termed as a *framework first* strategy.)

Some small empirical studies that investigated the extent to which either strategy was adopted, and with what level of success, are described in (Pohthong & Budgen 2000) and (Pohthong & Budgen 2001). These used a rather constrained component context (Unix processes), and as might have been expected, for much of the time the participants worked opportunistically, adjusting the choice of strategy as the solution evolved. However, for less experienced designers, the use of an *element first* strategy did appear to be more likely to result in a working solution. It was thought that this might be partly because identifying the available components might provide better assistance with creating a conceptual model of how the overall application might work.

Reusing existing components with an opportunistic strategy makes it possible to make good use of *exploratory* prototyping. When the option to develop components rather than simply reuse existing ones is added to this (what we might term a 'buy or build' strategy) then the design decisions become even

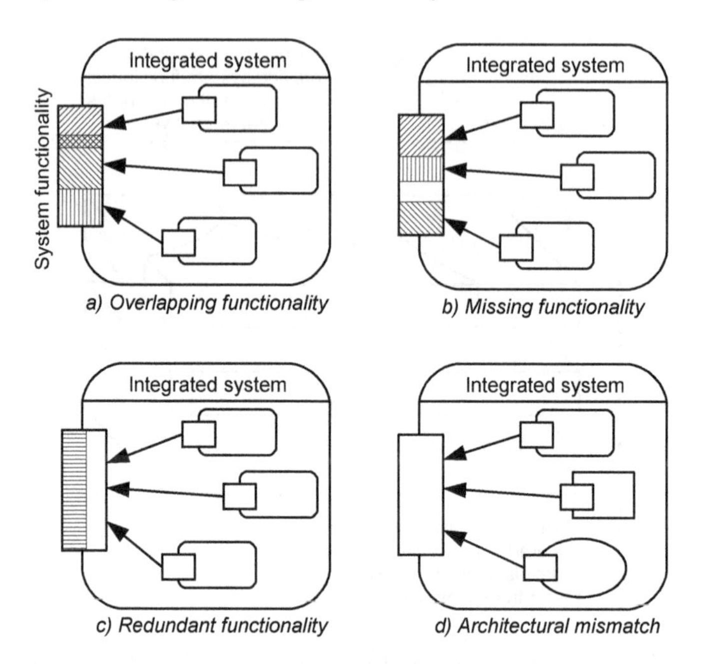

FIGURE 16.2: Illustrations of some of the problems that might arise in component integration

more complex. As indicated in Chapter 11, it is necessary to factor in the knowledge that the cost of developing components for reuse is around four times greater than direct development. Given these issues, we can conclude that simple strategies are unlikely to work for many applications, and that the complexity of component-based design can only really be addressed by an opportunistic strategy.

In composing a system from a set of components, particularly when working opportunistically, a designer needs to be able to model and predict their aggregate behaviour and functionality. In doing this, the designer also needs to identify the potential for the occurrence of any of the issues shown schematically in Figure 16.2. These problems can be classified as follows.

a) **Overlapping functionality.** This occurs where two or more components are able to perform a particular application function. The designer's task is to determine which one should perform the task, and how to ensure that only that component performs it. From a design point of view it is definitely undesirable to have the same function performed by different elements (possibly with different limits or ranges), and this also creates a problem for system maintenance and evolution.

b) **Missing functionality.** The problem arises when the total set of functions provided by the components chosen to make up the application is less than what is needed. The solution is fairly simple, either find another component or (possibly) create one. So this should be a short-term issue with no longer-term technical debt associated with it.

c) **Redundant functionality.** Components may well provide services over and above those that form the basis for choosing those particular components. This is particularly so for large components. The designer needs to choose between incorporating the added functionality in some way (with possible undesirable consequences) or, preferably, finding some way to exclude the unwanted functions.

d) **Architectural mismatch.** This issue was discussed in Chapter 11, and arises when there are mismatches between the expectations that each component makes about its context. The study by Yakimovich, Bieman & Basili (1999) provides examples of how architectural mismatch can arise from inconsistencies in:

- *component packaging*—where components may be constructed as *linkable* elements that need to be integrated into an executable image, or as *independent* elements, that may be scheduled independently;
- *type of control*—concerning the way that control is organised and transferred between components, and whether this is managed centrally within an application or concurrently (such as through an event-driven mechanism);
- *type of information flow*—whether organised as control flow through method calls, data flow through shared memory, or in a mixed format;
- *synchronisation between components*—referring to whether a component can 'block' the execution of other components (synchronous), or can continue to run regardless (asynchronous);
- *binding time*—referring to when the components are attached to connectors, which may occur at compilation, link-editing, run-time etc.

What should be clear from this is that while component-based development offers potential to bring together components from a variety of sources and in a range of forms, it is probably wisest to employ only components that conform to a single architectural style. This doesn't prevent the first three problems arising, but these are likely to be much more tractable than coping with architectural mismatch.

The systematic mapping study by Vale et al. (2016) identified some important gaps in CBSE knowledge, while recognising that some of these were compounded by the diversity of CBSE forms and application areas. Gaps of particular relevance to design included a lack of experience reports about use of CBSE in practice, a lack of clear concepts about CBSE in practice, and a lack of CBSE tool support. CBSE tools do exist but appear to be largely concerned with implementation rather with activities such as modelling.

To complete our discussion of designing with components we look at how components might be used to compose the software needed to support the business of the CCC.

Modular implementation for the CCC

Most of the operations of the CCC are those that will be common to many businesses that provide resources on a temporary basis, whether it be car rental, hire of scaffolding, use of hotel meeting rooms etc.
A CBSE implementation might therefore expect to find modules that can handle its needs as regards accounting for use, keeping records of cars and customers. (Of course, as we have already observed, it will be better if these all conform to the same component architecture.)
Building the user interface app, to be used on mobile phones, can again use fairly low-level components such as buttons, boxes etc., and one benefit of using a suitable component platform is that these should be portable across different platforms.
What is less likely to be 'off the shelf' is the rather important element of *reservation.* Locating nearby cars and calculating their distance from the customer and then ranking these in some way may be rather less readily available, although some of the elements needed for GPS location may well be so.
Whether to adopt this strategy also depends upon issues such as trust (of components built by others), confidence in their continued availability (and evolution). As with all issues associated with ISPs there is no one answer of course.

16.3 Designing with software services

With any new implementation technology paradigm, there is an inevitable lag between it becoming available and the accumulation of relevant *knowledge schema* about how to design applications using it. And for more complex forms, such as objects, components and services, there is also a need to establish a consistent context for their use, usually in the form of some sort of *framework.*

The basic service model, often described as *Software as a Service* separates *possession* and *ownership* of software from its *use* by employing remote service providers to deliver services 'on demand' (Turner et al. 2003, Budgen, Brereton & Turner 2004). It is this paradigm that underpins the concept of the *cloud,* whereby the user of a service needs to have no detailed knowledge of how it is provided. Indeed, the use of software services depends upon the use of *business modelling* at least as much as more 'technological' design modelling.

Designing for very basic use of the SaaS model then requires little more than using the pattern shown in Figure 15.13, possibly adding an element of orchestration to aid with selecting services.

Beyond that, the lack of clear definitions of what exactly is meant by such terms as *Service Oriented Architecture* becomes quite a problem. A systematic mapping study performed to identify what exactly was considered to comprise an SOA found that of 921 studies referring to SOA, only 98 provided explicit definitions of what the authors considered this to consist of (Anjum & Budgen 2012). Where definitions did exist, they differed in such aspects as the level of abstraction and the assumed context (consumer, provider, developer).

Integrating the terms that were used to describe SOA in those studies that did provide some form of definition produced a set of characteristics that could be considered to provide a profile of SOA (and which have no particular conflicts or inconsistencies). And hence, these can be considered as factors that need to be considered when designing an application that is to be realised using service forms, and that should ideally be provided by a service framework. (Of course, the application may itself be providing a service.) In brief, these characteristics are as follows.

- *Architecture.* This relates to the overall organisation of an application in terms of the ways that the different services interact. Implicitly, these interactions are usually through SOAP, although REST is also used.

- *Binding Time.* This refers to the time at which a particular service (and hence the service provider) is selected by an application. Some services may be bound when the application is composed, since their use may be intrinsic to its purpose, while others may well be bound dynamically on demand.

- *Capability.* This relates to the purpose of an application or service, viewed from the perspective of the service consumer. The emphasis here lies upon the *functional* viewpoint, as services are essentially *stateless* in nature, providing limited scope to use *behavioural* modelling in any way.

- *Composition.* This relates to the specific configuration of services employed to meet the needs of the end-user (often termed *orchestration*, particularly in a dynamic context). Composition is effectively a realisation of the *design model*.

- *Contracts.* Because the use of services is likely to involve third parties as suppliers, it is common for a service provider to offer a *Service-Level Agreement* that specifies the terms and conditions under which a service will be provided and used. And of course, this may have to be managed dynamically (one of the more challenging aspects of the service model) unless such contracts are negotiated in advance of use.

- *Delivery.* Essentially this is the operational aspect by which a service provides the agreed service functionality.
- *Distributed Sources.* Services can be delivered locally as well as from remote sources, but this aspect is particularly concerned with delivery across a network (implicitly, the web).
- *Identity.* This is essentially the description of a particular service and the ways that it can be accessed.
- *Interoperability.* This relates to the service framework and the way that services can be deployed within this, including the possibility that the end-user has no knowledge about the location or provider of a particular service, and indeed, that such a service may be provided by different sources each time it is requested.
- *Packaging.* A constructional issue concerning the creation of services so as to ensure that a service both provides the required functionality, and also the appropriate interface.

While all of these are factors that influence design of an application in some way, not all of them will necessarily be a direct influence. Characteristics such as *packaging* and *identity* are primarily ones that are determined or specified by the service provider, and hence likely to be an indirect influence. Others such as *orchestration*, *binding time* and *capability* are likely to be of much more immediate importance when configuring an application to use services. Figure 16.3 shows how these characteristics map on to the main elements of the service paradigm, by annotating the model provided in Figure 11.5.

Development of SOA applications is apt to employ a complex mix of business and technological modelling. As noted in the last chapter, the use of *design patterns* has been advocated for more detailed modelling needs (Erl 2009). There has also been some interest in using what are more or less plan-driven forms of approach such as the *SOA method* (SOAM) proposed in (Offermann & Bub 2009). (There is some discussion of different methods in (Anjum & Budgen 2017).) Modelling would appear to be largely dependent upon the use of UML class and component diagrams.

In many ways, designing SOA applications can be considered as being an indirect form of design. Rather than explicitly determining things like configuration of elements, the SOA designer is more concerned with writing a set of *rules* for the *dynamic orchestration* of the services. And these roles are not only concerned with basic service properties, but also with things like *Service-Level Agreements* that are confirmed or re-negotiated at run time, as well as *security*.

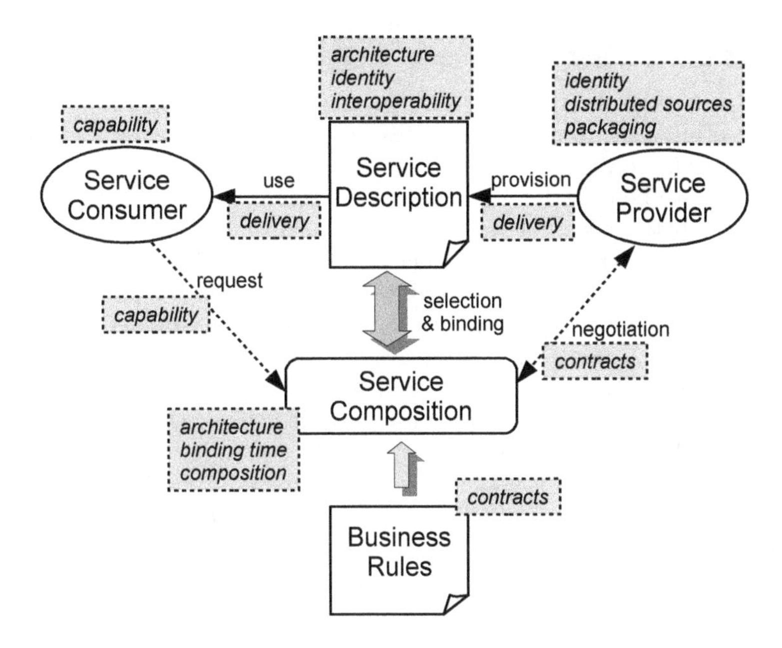

FIGURE 16.3: SOA characteristics mapped on to the service model

One question that arises when designing around the use of SOA is the extent to which relevant commercial service provision exists? (A similar issue arises with component-based design, but once a component has been found, its use is less dependent upon the external provider.) As with the use of physical services, adoption of this model does involve an element of *risk* in that a service provider may cease to trade. Hence employing a service model does implicitly require the availability of multiple providers for a given service. Many of the papers describing and discussing use of services do tend to be focused on quite specific application areas, which may help the emergence of a provider market, possibly related to the use of *cloud* services.

To complete our discussion of the use of software services in this section, we again turn to the needs of the CCC and consider how this might make use of a service model, including services provided by external providers. For the purposes of this discussion, we assume that some of the more 'general purpose' services needed are readily available from external sources.

Software service implementation for the CCC?

We can model the customer's use of the CCC as being provided by a set of distinct user-facing services: seeking available cars; making a reservation; accessing the chosen car. However, all, or most, of these will need to be provided by a 'bespoke' service provider since they are specific to the needs of the CCC. Of course it is also possible that this specialist service provision may be based within the CCC system. There is no reason why 'local' and 'general' services should not be used together as long as they conform to the same service model. (Making a reservation is more likely to be available as a general service.)
The backend functionality may well be able to use a number of services that are sufficiently widely used that it is likely that they can be provided by multiple service providers. This may include such tasks as identifying where the cars are in the streets (GPS locations), as well as for accounting and billing functions, with security clearly being important for the latter.

16.4 Empirical knowledge about modular design

The diversity and emerging nature of the technologies covered in this chapter pose something of a challenge for empirical studies. We have already identified the systematic mapping study by Vale et al. (2016) as looking at research activities in CBSE. At time of writing, there is no obvious equivalent study addressing the use of software service technologies.

Key take-home points about designing with components and services

If we view components and software services as being evolutionary stages in modular design of software, we can identify the following lessons about their use.

Design models. Modular design can make use of many of the modelling forms already developed for object-oriented development, possibly augmented by forms such as DFDs where the 'flow' nature of service models require such a form.

Design by composition. The emphasis placed upon *reuse* means that the main design activity is centred upon *composition* of pre-existing elements. This in turn changes many of the supporting activities, which now involve such tasks as finding modules, negotiating terms of use, and addressing the question of interfacing between modules from different sources.

Business models. One reason for using modular forms is to be able to rapidly assemble software to meet business needs, particularly where these may involve applying application-specific 'rules' to the way that fairly standard tasks are performed. So business modelling explicitly plays a much more important role in the use of these technologies than may be the case for other forms.

Chapter 17

How Good Is My Design?

After looking at different ways of producing and describing design ideas in the preceding chapters of Part III, this chapter now considers some of the ways that we can reflect upon and assess the *quality* of a particular design.

So, why do we need to do this? Well, at the very beginning of this book, in explaining why software development was an example of an *ill-structured problem*, and describing what this meant, there were two characteristics of ISPs that were described there that are particularly relevant to this chapter.

- There is no *ultimate test* for a solution to an ISP, by which the designer or design team can determine that they have produced a design model that fully meets the customer's needs.
- There is no *stopping rule* that enables the designer(s) to know that the current version of the design model is as good as it is going to get and that there is little point in refining it further.

These two characteristics are reflected in much of the material covered in the third part of this book. Whether considering plan-driven development, agile development, use of patterns or just plain opportunistic design activities, we have no means of knowing when our design activities should stop. And while we might cope with these characteristics by considering our goal to be one of *satisficing*, by aiming to produce a design that is 'good enough', there are no readily available measures that can tell us whether or not that has been achieved either.

This chapter therefore takes a brief look at three (related) issues. Firstly, we consider the question of what concepts and measures we might use to assess design *quality*. Secondly, we look at a technique that can be used to help make such an assessment. Thirdly, we briefly describe what might we then do if it is obvious that changes are needed.

It is also worth putting the material of this chapter into perspective with regard to the material covered earlier in Chapter 5. There the aspect of concern was more one of assessing what we might term the 'human and social' aspects of software design—that is, examining the ways that people design things, and the manner in which end-users deploy the products. This chapter is more concerned with the design models that form the outcomes of designing, and the quality measures that can be employed to assess their form and structure. Hence it has a more 'technology-centric' focus.

17.1 Quality assessment

The concept of *quality* is of course something of a subjective one. If a group of friends have enjoyed eating a curry together, they might still come out with quite different opinions of its quality. One who only occasionally eats Asian food might consider it to have been very good; another who greatly likes Asian food might think it was 'good but a bit bland'; while a third who doesn't really like spicy dishes may feel it wasn't to their taste. And of course, assuming that they are not actually eating this in Asia, yet another, who has extensive knowledge of Asian cookery, might consider it to have been rather below the standard to be expected in Asia. And at the same time, as all of the group will have eaten everything that was put before them, the curry can be considered to have *satisficed* their needs.

So concepts of quality can have a strong contextual aspect, and also be difficult to quantify (as in the example of the curry).

For software design too, the issue of quality can be difficult to quantify. One approach is to employ some quantitative measures (such as those used for object-oriented design that were described in Chapter 10) and then to *interpret* these within the context and needs of a particular software development project.

So, if we are to do that, our first question is to determine what constitutes characteristics that reflect 'good' quality in a design model? In Chapter 4 we identified some design *principles* that are generally considered to characterise a good design.

- *Fitness for purpose.* This comes as close as we can hope to providing the ultimate test for an ISP, but of course, is not something that we can easily quantify or indeed, assess systematically.
- *Separation of concerns.* This can be considered as providing an assessment of the 'modular' organisation of a design.

- *Minimum coupling.* For any application, the different elements need to interact in order to perform the necessary tasks, and coupling determines how they interact (statically and dynamically).
- *Maximum cohesion.* Like coupling, the idea of cohesion reflects our expectation that the compartmentalisation of the system elements provides elements that are relatively self-contained.
- *Information hiding.* Relates to knowledge used by the application and how this is shared between the elements, such that each element can only access the knowledge that is relevant to its role.

While these are important software design principles, they very much embody ideas about how a design model should be *organised*, strongly influenced by the idea that a design is likely to evolve and to be adapted.

Information hiding

A rather different perspective, and one that is equally important, is to think about the role and functioning of an application. The group of quality factors that we associate with this are often referred to as the *ilities* (because most of them end with 'il-ity'), and we briefly consider some of these below.

17.1.1 The 'ilities'

There are many *ilities*, with the emphasis placed upon particular factors being dependent upon the purpose of the application being developed. Here, the discussion is confined to a group that can be considered to be fairly widely applicable: reliability; efficiency; maintainability; and usability. Other, rather more specialised ones include: testability; portability; and reusability.

Reliability. This factor is essentially concerned with the dynamic characteristics of the eventual application, and so involves making predictions about behavioural issues. From a design perspective, assessing this involves determining whether the eventual application will be:

- *complete*, in the sense of being able to react correctly to all combinations of events and system states;
- *consistent*, in that its behaviour will be as expected and repeatable, regardless of how busy the application might be;
- *robust* when faced with component failure or similar conflicts (this is usually referred to as being *graceful degradation*)—for the example of the CCC, this might require that it copes with a situation where the customer is unable to unlock the selected car, despite having the correct code.

This factor is a key one for safety-critical systems, which may use replicated hardware and software to minimise the risk of component failure, but it is also important for those applications that have an element of user interaction.

Efficiency. This can be measured through an application's use of resources, particularly where these need to be shared. Depending upon context, the relevant resources may include processor time, storage, network use etc., and this multi-variate aspect does make it difficult to assess.

There are also trade-offs involved between these. As an example, economising on memory use might result in an increase on processor load. Where resources are constrained it can be important to try and make some assessment of the likely effects of particular design choices, although fine-tuning may be better left to the implementation phase.

Maintainability. The lifetime of a software artifact, whether it be a component or an application or a platform will probably be quite long. Planning for possible change is therefore something that should influence design decisions.

This factor strongly reflects the characteristic of *separation of concerns*. The problem is to determine what the likely evolutionary pathways may be. It may be important to be clear about the assumptions that an application makes about its context and use, and to query these in the light of different possible scenarios of evolution.

Usability. There are many things that can influence the usability of an artifact (Norman 2002). However, for software, the user-interaction, or HCI (Human Computer Interaction) elements tend to predominate and may well influence other design decisions too. The set of measures provided by the *cognitive dimensions* framework is widely considered as being a useful way of thinking about usability (Green & Petre 1996), and the concept of *interaction design* has helped focus attention on the nature of software use (Sharp et al. 2019).

The cognitive dimensions framework can be considered as providing a set of *qualitative* measures that can be used to assess a design (not just from the HCI aspect either). We now look at ways in which we might use *quantitative* measures, and the forms that they might take.

17.1.2 Design metrics

Unfortunately, neither the general design characteristics nor the 'ilities' can be measured directly, which is where the use of quantitative software *metrics* may assist by providing some surrogate measures. (Of course, we can use qualitative metrics too, particularly for assessing such factors as *usability*, but deriving values for these is apt to be quite a time-consuming task.)

We encountered some examples of metrics in Chapter 10 and this may be a useful place to examine their role and form a bit more fully. Measurement science differentiates between an *attribute*, which we associate with some element (in this case, of design), and a *metric*, which will have a value in the form of a specific number and an associated unit of measurement. And for a given metric, as we say in Chapter 10, we need to have some form of *counting rules* that are applied in order to derive its value. Indeed, Fenton & Bieman (2014) have observed that "measurement is concerned with capturing information about *attributes of entities*". So, our next need is to determine how ideas about quality can use metrics and measurements.

Figure 17.1 shows a simple view of the mapping that links quality concepts through to actual counts.

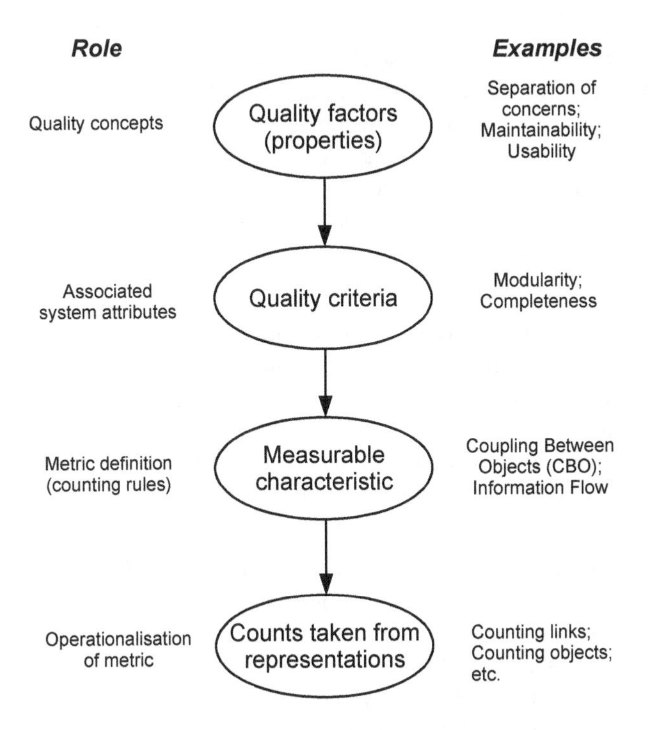

FIGURE 17.1: Linking quality concept to measurements

Unfortunately, none of the mappings between the elements of this are particularly easy to define. The desirable design characteristics are quite abstract, as are the attributes. Finding measures for an invisible and complex media such as software can be quite challenging, and so each of the things that we can actually measure usually comes down to counting tokens. For code metrics, the tokens are usually syntactic elements, for design metrics, they tend to be diagrammatical elements. So what we *can* count is not always what we would *like* to be able to count! And of course, we are limited to counts of static relationships, although we know that software has quite important dynamic properties.

So, counting the number of objects, components, sub-programs and the coupling links between these is probably the best we can do when looking at a design model. (We saw this in Chapter 10 when reviewing the Chidamber and Kemerer metrics.) Having relatively 'formal' models that are documented using (say) UML notations may help, but quite a lot can be done just by looking at sketches too. And there are other things that are related to the design model that can also be measured, such as the number of times the design of a particular sub-system of an application has been modified.

Before discussing what we might do with such knowledge, it is worth noting that there are two types of metric that are both commonly used in software engineering (Hughes 2000).

- *Actionable metrics* are those that relate to things we can *control*. We can react to the values obtained from using such a metric by making changes to our design. If we are counting the arguments of methods as part of a metric, we can make changes if we find a particular method seem to have too many arguments for its purpose. At a more abstract level, we can make changes if we find that a particular class has particularly high counts for coupling measures (when compared with the other classes in the design model).
- *Informational metrics* tell us about things that can be measured and that are important to us, but that we cannot influence directly. One example of this is 'module churn', which occurs where the number of revisions to the design (or code) of each module is counted over a period of time, and we observe that a particular element may have been modified many times. This churn is a consequence of design activities, but usually we wouldn't want to use it to control them!

Many design metrics fall into the category of actionable metrics, since after all, they are usually relating to an application that has yet to be built.

And as a last point about metrics, we should note that *complexity* is not an attribute in its own right. It is incorrect to refer to the 'complexity of object D', because there are many measures that might be applied to that object. Rather, complexity can be viewed as a threshold value for a particular measure—so that we consider anything above that value to be complex. As an example, we might argue that an object with more than (say) 12 external methods will potentially be 'complex' to use, because the developers may need to keep checking which methods are the ones they need.

And even then there are further pitfalls with the concept of complexity. Such a threshold value is not absolute, it may differ between applications and designs. And sometimes, a particular object will have a value that is above our chosen threshold value because of its role. The need to *interpret* the values of any metrics and decide whether or not they are appropriate, then leads us on to our next topic.

17.2 Reviews and walkthroughs

A technique that has proved itself useful in assessing design structure and likely behaviour is the design *review* or *walkthrough*, also sometimes termed an *inspection* although that term is more commonly used when assessing code rather than designs. The use of reviews dates from very early in the evolution of software engineering (Fagan 1976), and a set of basic rules for conducting design reviews has been assembled from experience (Yourdon & Constantine 1979, Weinberg & Freedman 1987, Parnas & Weiss 1987).

There are actually two forms of review that are in common use. The *technical review* is concerned with assessing the quality of a design, and is the form of interest to us in this section. The *management review* addresses such issues as project deadlines and schedule. Indeed, one of the challenges of review-driven agile approaches such as Scrum is keeping these issues distinct (and of course, they are not independent, which also needs to be recognised).

Technical reviews can include the use of forms of 'mental execution' of the design model with the aid of use cases and scenarios and so can help with assessing dynamic attributes as well as static ones. The sources cited above provide some valuable guidelines on how such reviews need to be conducted so as to meet their aims and not become diverted to issues that are more properly the domain of the management review. (One reason why these guidelines remain relevant is that they are really 'social' processes, and hence their effectiveness is uninfluenced by such factors as architectural style.) And it is important that the review does not become an assessment of the design team, rather than of the design itself.

Even with the input of metric values, a design review cannot provide any well-quantified measures of 'quality'. However, the review is where such concepts as complexity thresholds may be agreed (since there are no absolutes). With their aid it can help to identify weaknesses in a design, or aspects that might potentially form weaknesses as details are elaborated or when evolutionary changes occur in the future. It therefore provides a means of assessing potential *technical debt.* Of course we can rarely avoid having some technical debts; what matters is how much we have and the form it takes. In particular, if carefully planned and organised (and recorded) a review brings together those people who have both the domain knowledge and the technical knowledge to be able to make realistic projections from the available information about the design model.

Parnas & Weiss (1987) have suggested that there are eight requirements that a 'good' design should meet.

1. *Well structured:* being consistent with principles such as information-hiding and separation of concerns.

2. *Simple:* to the extent of being 'as simple as possible, but no simpler'.

3. *Efficient:* providing functions that can be computed using the available resources.

4. *Adequate:* sufficient to meet the stated requirements.

5. *Flexible:* being able to accommodate likely changes in the requirements, however these are likely to arise.

6. *Practical:* with module interfaces providing the required functionality, neither more nor less.

7. *Implementable:* with current (and chosen) software and hardware technologies.

8. *Standardised:* using well-defined and familiar notation for any documentation.

The last point is particularly relevant given that as we have observed, designers often employ informal notations. However, the key issue is that any notations used should at least be readily explained and interpreted. A set of hand-drawn sequence diagrams may be sufficient to explore scenarios without the need to follow exact UML syntax.

A systematic review looking at the use of metrics by industry during agile software development by Kupiainen, Mäntylä & Itkonen (2015) made some observations of how such reviews could use metrics. In particular the review reinforced the view that the targets of measurement should be the product and the process, but not the people. They also observed the following.

- Documentation produced by a project should not be the object of measurement, and the focus of attention should be the product and its features.

- Using metrics can be a positive motivator for a team, and can change the way that people behave by determining the set of issues that they pay attention to.

- For agile development, teams used metrics in a similar way to how they were employed with plan-driven projects, using them to support sprint and project planning, sprint and progress tracking, understanding and improving quality, and fixing problems with software processes.

An important ground-rule for a design review is that it should concentrate on *reviewing* the design and not attempt to fix any issues that are identified. In the next section we therefore look at what we might do next, following a review.

17.3 Refactoring of designs

The idea of *refactoring* was introduced in Section 14.7, and as explained in Section 15.4, the need to perform refactoring is often motivated by the presence of *code smells* (Fowler 1999). The discussion in both of these sections was focused largely upon code, rather than design, as this is often where the relevant structural weaknesses are identified. However, many of the code smells originate in design weaknesses and there is scope to identify their likely presence during a design review. Of course, this isn't always the case, and the idea of *code duplication* is an obvious example of a problem that is less likely to be identified during design.

Refactoring can occur before that though, based upon the design model, and is one way of addressing the issues identified in a design review. Clearly, some issues such as those associated with duplication are unlikely to be flushed out by a review. However, the need for redistribution of functionality across design elements (objects, components) and reorganisation of communication models are possible outcomes, along with the possible creation of new design elements.

What this points to is the need to *document* and record the outcomes of a review with care. One possible option here is to use video-recording, with the agreement of all participants, so that issues do not get lost and any whiteboard-based discussions are captured. And of course, after a post-review refactoring of the design (if such changes are necessary), it may be useful to repeat the review, or part of it, using the same review team.

17.4 Empirical knowledge about quality assessment

Empirical studies that are relevant to this chapter are chiefly those looking at the use of metrics in assessing quality as well as those examining the effectiveness of code reviews and ways of improving their efficiency. (Design reviews do not appear to have been very extensively studied, perhaps because of the difficulty of finding enough experienced participants.) Earlier we mentioned the systematic review by Kupiainen et al. (2015) and its findings. Here we identify a small number of further studies that use different approaches for the purpose of evaluation.

Software metrics can be used inappropriately as well as being a useful tool, and before adopting any metrics it may well be worth reading the book by Fenton & Bieman (2014) and the invited review paper by Kitchenham (2010). Another interesting study is that by Jabangwe, Borstler, Smite & Wohlin (2015) which looks at the link between object-oriented metrics and quality attributes. Reinforcing our observations about OO metrics in Chapter 10, this review did note that measures for complexity, cohesion, coupling and size corresponded better with reliability and maintainability than the inheritance measures.

The study reported in Bosu, Greiler & Bird (2015) is based on interviews with staff at Microsoft and provides some useful recommendations for improving the practice of reviews. The effect of using light-weight tool-based review practices is reviewed in McIntosh, Kamei, Adams & Hassan (2016) using three large open-source repositories in a case study design. Again, the findings of this study reinforce the value of well-conducted reviews.

Key take-home points about assessing design quality

Assessing design quality poses some challenges, not least because our interpretation of quality may be influenced by such factors as domain, technologies etc. However, there are some useful general points worth making here.

Quality measures. A design will have many different quality attributes, usually related to the *ilities*, and with the emphasis placed on each attribute being largely dependent upon the nature of the application. Metrics that directly measure features that are directly related to these attributes are generally impractical and it may be necessary for a team to decide on their own counting rules when making measurements related to a design model.

Design reviews. The use of formal or informal design reviews provides an opportunity to make project-specific interpretation of any metric values obtained, as well as to assess the general structure of a design model. It is important that the outcomes from a review are carefully recorded.

Design refactoring. This can be performed as part of the outcome from a review, but there is little guidance available about how to perform refactoring at the design level.

Chapter 18

And What About...

This last chapter provides a brief summary of some of the topics that this book does *not* cover. These are some quite important influences upon software design that either don't really fit into the narrative of this book, or that don't contribute quite enough to the theme of emerging software design knowledge to be included. And inevitably, there will also be some topics that don't appear here that others may feel should do.

The structure of this chapter is a little different too, in that the discussion of empirical knowledge is embedded in the sections, rather than appearing in a separate section. There is also no final 'take home' section, since the descriptions are really too short to merit this.

18.1 Open source software (OSS)

This has had occasional mention in the preceding chapters. One description for the way that *open source software* (OSS) is developed is that this is a *reactive* development process, whereby a particular application evolves largely in response to what its developers perceive as being needed. It can also be viewed as being a form of incremental development. Either way, it has become a major source of software that is used worldwide for an ever-increasing number of applications. For most users, the most important characteristic is usually that it is free to download and use, but for some the availability of the source code and the freedom to change it opens up important opportunities.

The ramifications of OSS can be somewhat theological in nature, but most would consider that its roots lie back in the pioneering work of Richard Stallman's *Free Software Foundation* and the GNU project (a recursive definition of GNU is 'GNU's Not Unix'!). Two factors accelerated its emergence in the 1990s: one was when the Linux operating system kernel came together with GNU to provide a widely distributed free operating system for personal computers; and the other was the emergence of the world-wide web as a platform for ready distribution and downloading of software applications. The establishment of the *sourceforge*[1] web site as a trusted platform for hosting and distributing open source software has also provided an important element of confidence for users.

Our interest here however, does not lie in the complexities and philosophies of OSS definitions, licensing etc., or in the ever-growing corpus of widely used applications, but in the way that such software is *developed.* A relatively early study by Wu & Lin (2001) suggested that while the development process usually adopted by OSS projects was incremental, it also took a form that was very different to the way that this type of development process was used in more 'conventional' development practices. In particular, they observed that projects could work in a great variety of ways, ranging from the use of highly democratic processes for making decisions through to a 'dictatorship' model.

The key characteristic that they identified lay in the organisation of the development process, and particularly the massive coordination effort required, rather than in the development steps themselves. Instead of the ideas from a small number of decision-makers (designers) directing the efforts of many developers, the open source model involves one or a team of co-ordinators drawing together the ideas of many participants. Inevitably, simple model does not come near to describing all of the variations. Many projects have begun with a single developer (as did Linux) and it was the later extension and evolution that was influenced by the many.

OSS development tends to make extensive use of code reviews, although because teams may well be globally distributed, these are not necessarily in the form described in Chapter 17. OSS projects can also demonstrate quite complex social structures that go well beyond the relatively simple forms outlined above. And joining such a project can be a complex process in terms of the group dynamics involved, as is illustrated by the systematic review reported by (Steinmacher et al. 2015).

With the potential to involve so many contributors, open source development requires quite comprehensive support for change management (examples of tools include *git* and *subversion*, but new ones continue to emerge). The comprehensive and detailed records related to open source projects readily available from these code management systems makes them a valuable resource for empirical studies. However, the use of OSS data may also form a source of bias for their findings where these relate to the processes involved in design and development.

[1]www.sourceforge.net

So, in terms of contribution to design knowledge, OSS has broadened our understanding of how to organise potentially large and distributed teams, as well as encouraging automation of change records. It has certainly demonstrated the benefits of having many pairs of eyes scrutinising design ideas too.

18.2 Formal description techniques (FDTs)

The notations discussed in the previous chapters have largely been both diagrammatical and *systematic*, with most of them lacking formal mathematical underpinning (the main exception being the statechart). And as noted, the advantage of diagrams is that they can be easily modified and can be used informally—reflecting the way that many designers work.

While informal syntax and semantics may be useful when exploring ways of addressing an ISP, the lack of rigour can present problems for such tasks as verification. And for design problems that fall into such categories as *safety-critical*, or that are concerned with financial transactions, an inability to formally and rigorously check for correctness and completeness is a significant disadvantage. Most of us, when travelling on an aircraft that is using 'fly by wire' technology to manage the control surfaces, would like to think that the software used for this purpose had undergone *very* thorough and comprehensive examination and assessment to ensure correct functioning in all situations!

From the early days of software engineering, researchers have therefore tried to harness the powers of mathematical notation to aid with modelling such activities as software specification, design and testing. Unfortunately, while mathematical rigour may be achievable when solving many forms of WSP, the characteristics of ISPs make it difficult to employ mathematical reasoning very widely, particularly in a context that exhibits the combined characteristics of ISPs and of software that we discussed in Chapter 1. So although the resulting *formal description techniques* (FDTs) are sometimes referred to as *formal methods*, in practice they tend to have very powerful representation parts while providing very little guidance on how to use them for design.

While FDTs can be used almost anywhere in the software development process and life-cycle, the roles for which they are particularly well suited are:

- specifying *system properties* for requirements specification (black box) ; and
- specifying the detailed form of a *design model* in the detailed design stages (white box).

In addition, their use does require some mathematical training and the amount of time needed for their use may create a substantial overhead, even with the support of tools. The variety of notations used, and their relative unfamiliarity, may also have made them less attractive to project managers. Incorporating ideas about software architecture has also proved to be challenging.

In a fairly early study of their use, Hall (1990) observed that, despite various limitations upon how and where they could be deployed, FDTs had been used on a reasonably large number of real projects, particularly those with safety-critical features. However, they also seemed to have been most successful when used on key elements of applications, rather than when being used to develop the complete application.

Little seems to have changed since then. FDTs have become part of the software engineer's toolbox, albeit a rather specialised part, to be used as and when appropriate. Researchers have also emphasised the use of 'lightweight' forms, seeking to use the power of mathematics where it can be most useful. However, there is little evidence that industry has embraced their use very widely, and indeed, they do sit awkwardly alongside ideas such as agile development.

This section provides a very brief introduction to one of the best-known *model-based* specification forms, the *Z* language (pronounced as 'zed'). When creating a model-based specification, the specifier uses various mathematical forms to construct a 'model' of the application and then uses this to help reason about its properties and behaviour.

Z was created by J-R Abrial in the Programming Research Group at the University of Oxford, underwent major development in the 1980s, becoming established in the 1990s. The text by Spivey (1998) is widely regarded as providing the authoritative definition, but many other, more introductory, texts are also available. An 'object' version of the formalism is described in Smith (2000).

Z exhibits an unfortunate characteristic that typifies many formal specification forms, which is the use of mathematical symbols that are not easily reproduced in everyday typefaces. (In fairness, most are quite easily drawn on a whiteboard.) The only real process involved in its use is one of *refinement.*

The basic *vocabulary* of Z is based upon three elements: sets, set operations and logic, each with their own notational features.

Sets. A *set* is a collection of elements (or set members), in which no sense of ordering of elements is involved. In Z, there are some pre-defined set types, such as $\mathbb{Z}$, the set of all integers; and $\mathbb{N}$, the set of all natural numbers (positive integers and zero). In addition, the user may specify their own set types, and for many problems is likely to want to do so. The convention is to employ all upper case italic letters for the identifier of a set type, with set members being described in text or enumerated.

Examples are:

$[CCCCARS]$the set of all cars available for use by CCC members

$[RUNWAY] ::== main \mid north \mid west$

The first example is self-evident, while the second tells us that an airport has three runways. As in this example, the variables are in lower case.

Set operations. These are largely as would be expected. Two that may be less familiar are the use of # that returns the number of elements in the set, and $\mathbb{P}$ for the *powerset* of a set (the set of all subsets of that set).

Logic. Operations and relations in a specification are usually expressed by using the standard operators of predicate logic. The logical operators can then be combined with the set elements and used to describe characteristic rules affecting a set, usually in the general form of:

$$declaration \mid predicate \bullet expression$$

where the *declaration* introduces the variables of concern to the rule; the *predicate* constrains their values; and the *expression* describes the resulting set of values.

A key component of Z is the *schema*. The role of this is to describe a system operation, and it is usually drawn as an open box that has three elements:

- the schema *name*;
- the *signature*, that introduces any variables and assigns them to set theoretic types;
- the *predicates* that form the *state invariants*, describing how the elements of the signature are related and constrain the operation, and that are described in terms of *preconditions* and *postconditions*.

A very simple example of a schema that is related to the activities of the CCC is provided below.

ReserveCar

$drivers, drivers' : \mathbb{P}\ CCCMEMBER$
$d? : CCCMEMBER$

$\#drivers < \#CCCCARS$
$d? \notin drivers$
$drivers' = drivers \cup \{d?\}$
$\#drivers' \leq \#CCCCARS$

Here, the elements used are:

- the *name* of the schema, which is *ReserveCar*;
- the *signature* that describes the 'before' and 'after' states of the set of drivers (CCC members who are currently using a CCC car), with the convention being that the primed identifier denotes the state of the set after the operation; while the variable *d?* is a variable used for an input that is required as part of the schema;
- the *predicate* describes the operations required to reserve a car, which are as follows:
  - firstly, the number of drivers should be fewer than the number of cars available (*precondition 1*);
  - the new driver should not already be driving a car (and hence is not in the set of existing drivers) (*precondition 2*);
  - the new set of drivers will be comprised of the original set together with the new driver (*postcondition 1*);
  - the number of drivers after the operation must now be either less than the number of available cars, or equal to it (*postcondition 2*)

One advantage of this type of formal specification is that we can readily reason about it, and clearly there are some shortcomings we can identify fairly quickly. In particular, the set of cars implicitly includes all of the cars available at the site. But of course, some may be unavailable for other reasons, such as being serviced, and so we should probably use a subset of 'active' cars, together with another schema that keeps that subset up to date.

From a design perspective, Z probably does not particularly assist with the problems of producing a design solution for an ISP, beyond helping to manage completeness and consistency. In many ways it can make more of a contribution to the issue of clarifying the needs that an application needs to meet (that is, the specification). One of the characteristics of an ISP is the lack of a definitive specification, and while Z might not be able to address all the aspects of that lack, it can help with clarification.

Few empirical studies appear to have been performed on the use of FDTs, particularly with regard to their use in industry. This in turn makes it difficult to assess what features of their use are particularly valued by those who adopt them. However, the survey of use reported in Woodcock, Larsen, Bicarregui & Fitzgerald (2009) does provide some interesting illustrations of the use of formal techniques in industry as well as reviewing the domains where there has been adoption. As might be expected, many of the examples do fall into what might be termed as domains requiring 'high-integrity software'.

18.3 Model driven engineering (MDE)

Model-driven engineering (MDE) shares some characteristics with the FDTs discussed in the previous section. In particular, it would seem that a condition for successful adoption may well be for it to contribute to part of a project, rather than be used for the whole of it.

In Part II of this book, we focused strongly on the issue of producing design models, whether these were relatively informal ones, largely based on sketches, or more formalised, such as those using the UML notations. MDE is focused on making use of such models for automating various tasks, such as the generation of code, but also such issues as evolution and testing.

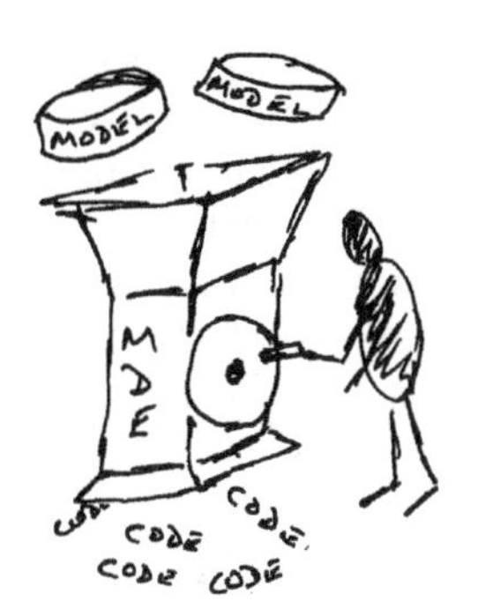

As noted by Schmidt (2006) MDE has built upon earlier technologies, such as CASE (Computer Assisted Software Engineering), and in the process has moved from using more generalised tools to more specialised ones such as domain-specific modelling languages (DSMLs), and generators that can create source code or simulations, observing that:

> "*instead of general-purpose notations that rarely express application domain concepts and design intent, DSMLs can be tailored via metamodeling to precisely match the domains semantics and syntax*"

MDE can also be employed with the UML as the modelling language, and the OMG (Object Management Group) provides and supports standards for this.

The use of MDE does appear to have been the subject of more critical analysis than FDTs, with examples such as the relatively early systematic review from Mohagheghi & Dehlen (2008), looking at the use of MDE in industry. At that point in time they did note the lack of suitably mature tools, and observed that they found "reports of improvements in software quality and of both productivity gains and losses, but these reports were mainly from small-scale studies".

A later survey by Whittle, Hutchinson & Rouncefield (2014) found quite widespread use of MDE, and observes that "many of the lessons point to the fact that social and organisational factors are at least as important in determining success as technical ones". They also note that: "the companies who successfully applied MDE largely did so by creating or using languages specifically developed for their domain, rather than using general-purpose languages such as UML". An interesting conclusion is that the real benefits to a company didn't stem from things like code generation but "in the support that MDE provides in documenting a good software architecture". One of the tips that they offer for successful use of MDE is to "*keep domains tight and narrow*", suggesting that it is easier to create DSMLs for "narrow, tight domains", rather than attempting to use MDE for broad areas of application. (There are some parallels with FDTs here, where success seems to have often been associated with quite narrow (but important) domains such as telecoms.)

And, where Moody (2009) criticises the visual aspects of the UML, from an MDE perspective Whittle et al. (2014) observe that:

> "*UML 2.0, for example, a major revision of the UML standard, didn't reflect the literature on empirical studies of software modelling or software design studies. Consequently, current approaches force developers and organisations to operate in a way that fits the approach instead of making the approach fit the people.*"

So from the perspective of this book, MDE does have some benefits when used appropriately, particularly with regard to helping with understanding and development of architectural features. Rather as FDTs require some familiarity with mathematical formalisms, successful use of MDE does need domain knowledge, but it is interesting to note that Whittle et al. (2014) also observe that the models themselves need not necessarily be formal ones, at least in the early stages of developing a design.

18.4 And the rest...

This book began with a discussion of the challenges posed by ill-structured problems such as those encountered during software development. The focus throughout this edition has been on the contribution that different techniques and forms make to addressing these challenges. And not only do ISPs have no *stopping rule*, a book like this doesn't either. There are many more approaches to modelling and designing software than those covered in the existing chapters, and many variations that arise too.

This last chapter has looked briefly at a small selection of some of the other techniques that software engineers have devised for modelling and designing software, over and above those covered in the main chapters. If your favourite technique has been omitted or overlooked, my apologies.

And of course, vast numbers of software systems are developed without making use of formal modelling or of any form of systematic design process (not that these are always developed well of course). What matters though is that anyone seeking to design software applications should have a good understanding of the medium and of the purposes of the application they are developing. And acquiring the relevant *knowledge schema* needs an understanding of the topics covered here, regardless of how they are eventually employed...

Bibliography

Adelson, B. & Soloway, E. (1985), 'The role of domain experience in software design', *IEEE Transactions on Software Engineering* **11**(11), 1351–1360.

Akin, O. (1990), 'Necessary conditions for design expertise and creativity', *Design Studies* **11**(2), 107–113.

Alexander, C., Ishikawa, S., Silverstein, M., Jacobson, M., Fiksdahl-King, I. & Angel, S. (1977), *A Pattern Language*, Oxford University Press.

Ali, M. S., Babar, M. A., Chen, L. & Stol, K.-J. (2010), 'A systematic review of comparative evidence of aspect-oriented programming', *Information and Software Technology* **52**(9), 871 – 887.

Allman, E. (2012), 'Managing technical debt', *Commun. ACM* **55**(5), 50–55. **URL:** *http://doi.acm.org/10.1145/2160718.2160733*

Anjum, M. & Budgen, D. (2012), A mapping study of the definitions used for Service Oriented Architecture, *in* 'Proceedings of EASE 2012', IET Press, pp. 57–61.

Anjum, M. & Budgen, D. (2017), 'An investigation of modelling and design for software service applications', *PLoS ONE* **12**(5), e0176936.

Arlow, J. & Neustadt, I. (2005), *UML 2 and the Unified Process: Practical Object-Oriented Analysis and Design*, 2nd edn, Addison-Wesley.

Avison, D. E. & Fitzgerald, G. (1995), *Information Systems Development: Methodologies, Techniques and Tools*, 2nd edn, McGraw-Hill.

Bailey, J., Budgen, D., Turner, M., Kitchenham, B., Brereton, P. & Linkman, S. (2007), Evidence relating to Object-Oriented software design: A survey, *in* 'Proceedings of First International Symposium on Empirical Software Engineering and Measurement (ESEM)', pp. 482–484.

Bano, M. & Zowghi, D. (2015), 'A systematic review on the relationship between user involvement and system success', *Information & Software Technology* **58**(148-169).

Barrow, P. D. M. & Mayhew, P. J. (2000), 'Investigating principles of stakeholder evaluation in a modern (is) development approach', *Journal of Systems & Software* **52**(2-3), 95–103.

Bass, L., Clements, P. & Kazman, R. (2013), *Software Architecture in Practice*, 3rd edn, Pearson.

Bates, B., Sierra, K., Freeman, E. & Robson, E. (2009), *Head First Design Patterns*, O'Reilly.

Beck, K. (2000), *Extreme Programming Explained: Embrace Change*, Addison-Wesley.

Beck, K. & Cunningham, W. (1989), 'A laboratory for teaching object-oriented thinking', *ACM SIGPLAN Notices* **24**(10), 1–6.

Bennett, K. H. & Rajlich, V. T. (2000), Software maintenance and evolution: A roadmap, *in* 'Proceedings of the Conference on the Future of Software Engineering', ICSE '00, ACM, New York, NY, USA, pp. 73–87.
URL: *http://doi.acm.org/10.1145/336512.336534*

Bennett, K., Layzell, P., Budgen, D., Brereton, P., Macaulay, L. & Munro, M. (2000), Service-based software: The future for flexible software, *in* 'Proceedings of Seventh Asia-Pacific Software Engineering Conference', IEEE Computer Society Press, pp. 214–221.

Blackwell, A. & Green, T. (2003), Notational systems—the cognitive dimensions of notations framework, *in* J. M. Carroll, ed., 'HCI Models, Theories, and Frameworks: Toward a Multidisciplinary Science', Morgan Kaufman, chapter 5.

Boehm, B. W. (1981), *Software Engineering Economics*, Prentice-Hall.

Boehm, B. W. (1988), 'A spiral model of software development and enhancement', *IEEE Computer* **21**(5), 61–72.

Boehm, B. W. (2002), 'Get ready for agile methods, with care', *IEEE Computer* **35**(1).

Booch, G. (1994), *Object-Oriented Analysis and Design with Applications*, 2nd edn, Benjamin Cummings.

Bosu, A., Greiler, M. & Bird, C. (2015), Characteristics of useful code reviews: An empirical study at Microsoft, *in* '12th IEEE/ACM Working Conference on Mining Software Repositories', IEEE Computer Society Press, pp. 1–11.

Breivold, H. P., Crnkovic, I. & Larsson, M. (2012), 'A systematic review of software architecture evolution research', *Information & Software Technology* **54**, 16–40.

Brereton, P. & Budgen, D. (2000), 'Component based systems: A classification of issues', *IEEE Computer* **33**(11), 54–62.

Brooks, F. P. (1987), 'No silver bullet: Essences and accidents of software engineering', *IEEE Computer* **20**(4), 10–19.

Brooks, F. P. (1988), Grasping reality through illusion – interactive graphics serving science, *in* 'Proceedings of the ACM SIGCHI Conference', pp. 1–11.

Brown, A. W. (2000), *Large-Scale Component-Based Development*, Prentice-Hall.

Brown, A. W. & Short, K. (1997), On components and objects: the foundations of component-based development, *in* 'Proceedings of 5th International Symposium on Assessment of Software Tools and Technologies', IEEE Computer Society Press, pp. 112–121.

Brown, A. W. & Wallnau, K. C. (1998), 'The current state of cbse', *IEEE Software* **15**(5), 37–46.

Brown, W. J., Malveau, R. C., McCormick, H. W. & Mowbray, T. J. (1998), *Antipatterns: Refactoring Software, Architectures, and Projects in Crisis*, Wiley.

Budgen, D. (2014), *Software Designers in Action: A Human-Centric Look at Design Work*, Chapman & Hall, chapter 12.

Budgen, D., Brereton, P., Drummond, S. & Williams, N. (2018), 'Reporting systematic reviews: Some lessons from a tertiary study', *Information and Software Technology* **95**, 62 – 74.

Budgen, D., Brereton, P. & Turner, M. (2004), Codifying a Service Architectural Style, *in* 'Proceedings 28th International Computer Software & Applications Conference - COMPSAC', IEEE Computer Society Press, pp. 16–22.

Budgen, D., Brereton, P., Williams, N. & Drummond, S. (2018), 'The contribution that empirical studies performed in industry make to the findings of systematic reviews: A tertiary study', *Information and Software Technology* **94**, 234 – 244.

Budgen, D., Brereton, P., Williams, N. & Drummond, S. (2020), 'What support do systematic reviews provide for evidence-informed teaching about software engineering practice?', *e-Informatica* **14**(1), 7–60.

Budgen, D., Burn, A., Brereton, P., Kitchenham, B. & Pretorius, R. (2011), 'Empirical evidence about the UML: A systematic literature review', *Software — Practice and Experience* **41**(4), 363–392.

Buschmann, F., Meunier, R., Rohnert, H., Sommerlad, P. & Stal, M. (1996), *Pattern-Oriented Software Architecture*, Wiley.

Carney, D. & Long, F. (2000), 'What do you mean by COTS? Finally, a useful answer', *IEEE Software* **17**(2), 83–86.

Chen, J., Xiao, J., Wang, Q., Osterweil, L. J. & Li, M. (2016), 'Perspectives on refactoring planning and practice: An empirical study', *Empirical Software Engineering* **21**, 1397–1436.

Chen, P. P. (1976), 'The entity-relationship model: Toward a unified view of data', *ACM Transactions on Database Systems* **1**(1), 9–37.

Chidamber, S. R. & Kemerer, C. F. (1994), 'A metrics suite for object oriented design', *IEEE Transactions on Software Engineering* **20**(6), 476–493.

Coleman, D., Arnold, P., Bodoff, S., Dollin, C., Gilchrist, H., Hayes, F. & Jeremes, P. (1994), *Object-Oriented Development: The Fusion Method*, Prentice-Hall.

Connor, D. (1985), *Information System Specification and Design Road Map*, Prentice-Hall.

Cooke, P. (1984), 'Electronic design warning', *Engineering Design* **9**(6), 8.

Coplien, J. O. (1997), 'Idioms and patterns as architectural literature', *IEEE Software* **14**(1), 36–42.

Crnkovic, I. & Larsson, M. (2002), 'Challenges of component-based development', *Journal of Systems & Software* **61**(3), 201–212.

Crnkovic, I., Stafford, J. & Szyperski, C. (2011), 'Software components beyond programming: From routines to services', *IEEE Software* **28**(3), 22–26.

Cunningham, W. (1992), The WyCash portfolio management system, *in* 'Addendum to the Proceedings on Object-oriented Programming Systems, Languages, and Applications (Addendum)', OOPSLA '92, ACM, New York, NY, USA, pp. 29–30.

Curtis, B., Krasner, H. & Iscoe, N. (1988), 'A field study of the software design process for large systems', *Communications of the ACM* **31**(11), 1268–1287.

Curtis, B. & Walz, D. (1990), The psychology of programming in the large: Team and organizational behaviour, *in* 'Psychology of Programming', Academic Press, pp. 253–270.

Cusumano, M. A. & Selby, R. W. (1989), 'How Microsoft builds software', *Communications of the ACM* **40**(6), 53–61.

De Marco, T. (1978), *Structured Analysis and System Specification*, Yourdon Press.

Détienne, F. (2002), *Software Design – Cognitive Aspects*, Springer Practitioner Series.

Diebold, P. & Dahlem, M. (2014), Agile practices in practice—a mapping study, *in* 'Proceedings 18th Conference on Evaluation & Assessment in Software Engineering (EASE 2017)', ACM Press.

Dybå, T. & Dingsøyr, T. (2008), 'Empirical studies of agile software development: A systematic review', *Information & Software Technology* **50**, 833–859.

Edwards, H. M., Thompson, J. B. & Smith, P. (1989), 'Results of survey of use of SSADM in commercial and government sectors in United Kingdom', *Information & Software Technology* **31**(1), 21–28.

Elbanna, A. & Sarker, S. (2016), 'The risks of agile software development: Learning from adopters', *IEEE Software* **33**(5), 72–79.

Ericsson, K. & Simon, H. (1993), *Protocol Analysis: Verbal Reports as Data*, MIT Press.

Erl, T. (2009), *SOA Design Patterns*, Prentice Hall.

Fagan, M. (1976), 'Advances in software inspections', *IEEE Transactions on Software Engineering* **12**(7), 744–751.

Fayad, M. E. & Schmidt, D. E. (1997), 'Object-oriented application frameworks', *Communications of the ACM* **40**(10), 32–38.

Feitosa, D., Avgeriou, P., Ampatzoglou, A. & Nakagawa, E. Y. (2017), The evolution of design pattern grime: An industrial case study, *in* 'Proceedings PROFES 2017 International Conference on Product-Focused Software Process Improvement'.

Fenton, N. E. & Bieman, J. M. (2014), *Software Metrics: A Rigorous and Practical Approach*, CRC Press.

Fichman, R. G. & Kemerer, C. F. (1997), 'Object technology and reuse: Lessons from early adopters', *IEEE Computer* **30**(10), 47–59.

Fink, A. (2003), *The Survey Handbook*, 2nd edn, Sage Books. Volume 1 of the Survey Kit.

Finkelstein, A., Kramer, J., Nuseibeh, B., Finkelstein, L. & Goedicke, M. (1992), 'Viewpoints: A framework for integrating multiple perspectives in system development', *Int. Journal of Software Eng. and Knowledge Eng.* **2**(1), 31–57.

Floyd, C. (1984), A systematic look at prototyping, *in* R. Budde, K. Kuhlenkamp, I. Mathiassen & H. Zullighoven, eds, 'Approaches to Prototyping', Springer-Verlag, pp. 1–18.

Fowler, M. (1999), *Refactoring: Improving the Design of Existing Code*, Addison-Wesley.

Gamma, E., Helm, R., Johnson, R. & Vlissides, J. (1995), *Design Patterns: Elements of Reusable Object-Oriented Software*, Addison-Wesley.

Gane, C. & Sarsen, T. (1979), *Structured Systems Analysis: Tools and Techniques*, Prentice-Hall.

Garlan, D., Allan, R. & Ockerbloom, J. (2009), 'Architectural mismatch: Why reuse is *still* so hard', *IEEE Software* **26**(4), 66–69.

Garlan, D., Allen, R. & Ockerbloom, J. (1995), 'Architectural mismatch: Why reuse is so hard', *IEEE Software* **12**(6), 17–26.

Garlan, D. & Perry, D. E. (1995), 'Introduction to the Special Issue on Software Architecture', *IEEE Transactions on Software Engineering* **21**, 269–274.

Green, T. R. G. & Petre, M. (1996), 'Usability analysis of visual programming environments: A 'cognitive dimensions' framework', *Journal of Visual Languages and Computing* **7**, 131–174.

Guindon, R. & Curtis, B. (1988), Control of cognitive processes during software design: What tools are needed?, *in* 'Proceedings of CHI'88', ACM Press, pp. 263–268.

Hadhrawi, M., Blackwell, A. & Church, L. (2017), A systematic literature review of cognitive dimensions, *in* L. Church & F. Hermans, eds, 'Proceedings of 28th PPIG Annual Workshop', Psychology of Programming Interest Group, pp. 1–12.

Hall, A. (1990), 'Seven myths of formal methods', *IEEE Software* **7**(5), 11–19.

Hannay, J., Dybå, T., Arisholm, E. & Sjøberg, D. (2009), 'The effectiveness of pair programming. A meta-analysis', *Information & Software Technology* **51**(7), 1110–1122.

Harel, D. (1987), 'Statecharts: A visual formalism for complex systems', *Science of Computer Programming* **8**, 231–274.

Harel, D. (1988), 'On visual formalisms', *Communications of the ACM* **31**(5), 514–530.

Harel, D. & Gery, E. (1997), 'Executable object modelling with statecharts', *IEEE Computer* **30**(7), 31–42.

Hatley, D. J. & Pirbhai, I. (1988), *Strategies for Real-Time System Specification*, Dorset House.

Hayes-Roth, B. & Hayes-Roth, F. (1979), 'A cognitive model of planning', *Cognitive Science* **3**(4), 275–310.

Heineman, G. T. & Councill, W. T. (2001), *Component-Based Software Engineering: Putting the Pieces Together*, Addison-Wesley.

Hughes, B. (2000), *Practical Software Measurement*, McGraw Hill.

Izurieta, C. & Bieman, J. M. (2007), How software designs decay: A pilot study of pattern evolution, *in* 'Proceedings of Empirical Software Engineering & Measurement, 2007'.

Jabangwe, R., Borstler, J., Smite, D. & Wohlin, C. (2015), 'Empirical evidence on the link between object-oriented measures and external quality attributes: A systematic literature review', *Empirical Software Engineering* **20**, 640–693.

Jackson, M. A. (1975), *Principles of Program Design*, Academic Press.

Jacobson, I., Booch, G. & Rumbaugh, J. (1999), *The Unified Software Development Process*, Addison-Wesley.

Jacobson, I., Christerson, M., Jonsson, P. & Overgaard, G. (1992), *Object-Oriented Software Engineering: A Use Case Driven Approach*, Addison-Wesley.

Jacobson, I., Spence, I. & Kerr, B. (2016), 'Use-Case 2.0', *Communications of the ACM* **59**(5), 61–69.

Jacobson, I., Spence, I. & Seidewitz, E. (2016), 'Industrial-scale agile—from craft to engineering', *Communications of the ACM* **59**(12), 63–71.

Jézéquel, J.-M. & Meyer, B. (1997), 'Design by contract: The lessons of Ariane', *IEEE Computer* **30**(1), 129–130.

Johnson, R. A. & Hardgrave, W. C. (1999), 'Object-oriented methods: Current practices and attributes', *Journal of Systems & Software* **48**(1), 5–12.

Jones, J. C. (1970), *Design Methods: Seeds of Human Futures*, Wiley-Interscience.

Kitchenham, B. (2010), 'What's up with software metrics? – a preliminary mapping study', *Journal of Systems & Software* **83**, 37–51.

Kitchenham, B. A., Budgen, D. & Brereton, P. (2015), *Evidence-Based Software Engineering and Systematic Reviews*, Innovations in Software Engineering and Software Development, CRC Press.

Kruchten, P. (2004), *The Rational Unified Process: An Introduction*, 3rd edn, Addison-Wesley.

Kruchten, P. B. (1994), 'The 4+1 view model of architecture', *IEEE Software* **12**(6), 42–50.

Kupiainen, E., Mäntylä, M. V. & Itkonen, J. (2015), 'Using metrics in agile and lean software development – a systematic literature review of industrial studies', *Information & Software Technology* **62**, 143–163.

Lanza, M. & Marinescu, R. (2006), *Object-Oriented Metrics in Practice*, Springer.

Larman, C. (2004), *Applying UML and Patterns: An Introduction to Object-Oriented Analysis and Design and Iterative Development*, 3rd edn, Prentice Hall.

le Goues, C., Jaspan, C., Ozkaya, I., Shaw, M. & Stolee, K. T. (2018), 'Bridging the gap: From research to practical advice', *IEEE Software* **35**(5), 50–57.

Littman, D. C., Pinto, J., Letovsky, S. & Soloway, E. (1987), 'Mental models and software maintenance', *Journal of Systems & Software* **7**, 351–355.

Litz, M. & Montazeri, B. (1996), 'Chidamber and kemerer's metrics suite: A measurement theory perspective', *IEEE Transactions on Software Engineering* **22**(4), 267–271.

Long, J. (2001), 'Software reuse antipatterns', *ACM Software Engineering Notes* **26**(4), 68–76.

Longworth, G. (1992), *Introducing SSADM Version 4*, Blackwell Publishing.

Lunn, K. (2003), *Software Development with UML*, Palgrave Macmillan.

Mangano, N., Toza, T. D. L., Petre, M. & van der Hoek, A. (2015), 'How software designers interact with sketches at the whiteboard', *IEEE Transactions on Software Engineering* **41**(2), 135–156.

Mayvan, B. B., Rasoolzadegan, A. & Yazdi, Z. G. (2017), 'The state of the art on design patterns: A systematic mapping of the literature', *Journal of Systems & Software* **125**, 93–118.

McIntosh, S., Kamei, Y., Adams, B. & Hassan, A. E. (2016), 'An empirical study of the impact of modern code review practices on software quallity', *Empirical Software Engineering* **21**, 2145–2189.

Miller, G. A. (1956), 'The magical number seven, plus or minus two: Some limits on our capacity for processing information', *The Psychological Review* **63**(2), 81–97.

Mills, H. (1971), Chief programmer teams, principles, and procedures, Technical report, IBM Federal Systems Division.

Mohagheghi, P. & Conradi, R. (2007), 'Quality, productivity and economic benefits of software reuse: A review of industrial studies', *Empirical Software Engineering* **12**, 471–516.

Mohagheghi, P. & Dehlen, V. (2008), Where is the proof? – a review of experiences from applying MDE in industry, *in* 'ECMDA-FA, LNCS 5095', Springer, pp. 432–443.

Moody, D. L. (2009), 'The "physics" of notations: Toward a scientific basis for constructing visual notations in software engineering', *IEEE Transactions on Software Engineering* **35**(6), 756–779.

Myers, G. J. (1973), 'Characteristics of composite design', *Datamation* **19**(9), 100–102.

Norman, D. A. (2002), *The Design of Everyday Things*, Basic Books.

Offermann, P. & Bub, U. (2009), Empirical comparison of methods for information systems development according to SOA, *in* 'Proceedings 17th European Conference on Information Systems (ECIS 2009)', pp. 1–13.

Olbrich, S. M., Cruzes, D. S. & Sjøberg, D. I. K. (2010), Are all code smells harmful? A study of god classes and brain classes in the evolution of three open source systems, *in* 'Proceedings of 26th International Conference on Software Maintenance', IEEE Computer Society Press.

Owen, S., Budgen, D. & Brereton, P. (2006), 'Protocol analysis: A neglected practice', *Communications of the ACM* **49**, 117–122.

Ozkaya, M. (2017), 'Do the informal and formal software modeling notations satisfy practitioners for software architecture modeling', *Information & Software Technology* **95**, 15–33.

Page-Jones, M. (1988), *The Practical Guide to Structured Systems Design*, 2nd edn, Prentice-Hall.

Pahl, G. & Beitz, W. (1996), *Engineering Design: A Systematic Approach*, 2nd edn, Springer-Verlag.

Palomba, F., Bavota, G., Penta, M. D., Fasano, F., Oliveto, R. & Lucia, A. D. (2018), 'On the diffuseness and the impact on maintainability of code smells: A large scale empirical investigation', *Empirical Software Engineering* **23**, 1188–1221.

Parnas, D. L. (1972), 'On the criteria to be used in decomposing systems into modules', *Communications of the ACM* **15**(12), 1053–1058.

Parnas, D. L. (1979), 'Designing software for ease of extension and contraction', *IEEE Transactions on Software Engineering* **5**(2), 128–137.

Parnas, D. L. & Clements, P. C. (1986), 'A rational design process: How and why to fake it', *IEEE Transactions on Software Engineering* **12**(2), 251–257.

Parnas, D. L. & Weiss, D. M. (1987), 'Active design reviews: Principles and practices', *Journal of Systems & Software* **7**, 259–265.

Perry, D. E. & Wolf, A. L. (1992), 'Foundations for the study of software architecture', *ACM Software Engineering Notes* **17**(4), 40–52.

Peters, L. & Tripp, L. (1976), 'Is software design 'wicked' too?', *Datamation* **22**(5), 127.

Petre, M. (2013), UML in practice, *in* 'Proceedings of the 2013 International Conference on Software Engineering (ICSE)', IEEE Computer Society Press, pp. 722–731.

Petre, M. & van der Hoek, A. (2016), *Software Design Decoded: 66 Ways Experts Think*, The MIT Press.

Pfleeger, S. L. & Atlee, J. M. (2010), *Software Engineering: Theory and Practice*, 4th edn, Pearson.

Pohthong, A. & Budgen, D. (2000), Accessing software component documentation during design: An observational study, *in* 'Proceedings of Seventh Asia-Pacific Software Engineering Conference', IEEE Computer Society Press, pp. 196–203.

Pohthong, A. & Budgen, D. (2001), 'Reuse strategies in software development: An empirical study', *Information & Software Technology* **43**(9), 561–575.

Pugh, S. (1991), *Total Design: Integrated Methods for Successful Product Engineering*, Addison-Wesley.

Purchase, H. C., Welland, R., McGill, M. & Colpoys, L. (2004), 'Comprehension of diagram syntax: An empirical study of entity relationship notations', *International Journal of Human-Computer Studies* **61**(2), 187–203.

Qumer, A. & Henderson-Sellers, B. (2008), 'An evaluation of the degree of agility in six agile methods and its applicability for method engineering', *Information & Software Technology* **50**, 280–295.

Radjenović, D., Heričko, M., Torkar, R. & Živkovič, A. (2013), 'Software fault prediction metrics: A systematic literature review', *Information & Software Technology* **55**, 1397–1418.

Ratcliffe, M. & Budgen, D. (2001), 'The application of use case definitions in system design specification', *Information & Software Technology* **43**(6), 365–386.

Ratcliffe, M. & Budgen, D. (2005), 'The application of use cases in systems analysis and design specification', *Information & Software Technology* **47**, 623–641.

Rentsch, T. (1982), 'Object-oriented programming', *ACM SIGPLAN Notices* **17**(9), 51–57.

Rittel, H. & Webber, M. (1984), Planning problems are wicked problems, *in* N. Cross, ed., 'Developments in Design Methodology', Wiley, pp. 135–144.

Robinson, P. J. (1992), *Hierarchical Object-Oriented Design*, Prentice-Hall.

Rumbaugh, J., Jacobson, I. & Booch, G. (1999), *The Unified Modeling Language Reference Manual*, Addison-Wesley.

Runeson, P., Höst, M., Rainer, A. & Regnell, B. (2012), *Case Study Research in Software Engineering: Guidelines and Examples*, Wiley.

Ruparelia, N. (2010), 'Software Development Lifecycle Models', *ACM SigSoft Software Engineering Notes* **35**(3), 8–13.

Schmidt, D. C. (2006), 'Model-driven engineering', *IEEE Computer* **39**(2), 25–31.

Schwaber, K. (2004), *Agile Project Management with Scrum*, Microsoft Press.

Schwaber, K. & Beedle, M. (2002), *Agile Software Development with Scrum*, Prentice Hall.

Shadish, W., Cook, T. & Campbell, D. (2002), *Experimental and Quasi-Experimental Design for Generalized Causal Inference*, Houghton Mifflin Co.

Shahin, M., Liang, P. & Babar, M. A. (2014), 'A systematic review of software architecture visualization techniques', *Information & Software Technology* **94**, 161–185.

Shapiro, S. (1997), 'Splitting the difference: The historical necessity of synthesis in software engineering', *IEEE Annals of the History of Computing* **19**(1), 20–54.

Sharp, H., Preece, J. & Rogers, Y. (2019), *Interaction Design: Beyond Human-Computer Interaction*, 5th edn, John Wiley & Sons.

Shaw, M. & Clements, P. C. (1997), A field guide to boxology: Preliminary classification of architectural styles for software systems, *in* 'Proceedings COMPSAC'97', IEEE Computer Society Press, pp. 6–13.

Shaw, M. & Garlan, D. (1996), *Software Architecture: Perspectives on an Emerging Discipline*, Prentice Hall.

Sheetz, S. D. & Tegarden, D. P. (1996), 'Perceptual complexity of object oriented systems: A student view', *Object Oriented Systems* **3**(4), 165–195.

Simon, H. A. (1973), 'The structure of ill-structured problems', *Artificial Intelligence* **4**, 181–201.

Sjøberg, D., Hannay, J., Hansen, O., Kampenes, V., Karahasanovic, A., Liborg, N. & Rekdal, A. (2005), 'A survey of controlled experiments in software engineering', *IEEE Transactions on Software Engineering* **31**(9), 733–753.

Smith, G. (2000), *The Object-Z Specification Language*, Springer.

Snyder, A. (1993), 'The essence of objects: Concepts and terms', *IEEE Software* **10**(1), 31–42.

Spivey, J. M. (1998), The Z notation: A reference manual, Technical report, Programming Research Group, University of Oxford.

Stapleton, J., ed. (1999), *Dynamic Systems Development Method (Version 3)*, Tesseract Publishing.

Stapleton, J., ed. (2002), *DSDM: Business Focused Development*, Addison-Wesley.

Steinmacher, I., Silva, M. A. G., Gerosa, M. A. & Redmiles, D. F. (2015), 'A systematic literature review on the barriers faced by newcomers to open source software projects', *Information & Software Technology* **59**(67-85).

Stevens, W. P. (1991), *Software Design: Concepts and Methods*, Prentice-Hall.

Stevens, W. P., Myers, G. J. & Constantine, L. L. (1974), 'Structured design', *IBM Systems Journal* **13**(115-139).

Szyperski, C. (1998), *Component Software: Beyond Object-Oriented Programming*, Addison-Wesley.

Taivalsaari, A. (1993), 'On the notion of object', *Journal of Systems & Software* **21**(1), 3–16.

Tempero, E., Gorschek, T. & Angelis, L. (2017), 'Barriers to refactoring', *Communications of the ACM* **60**(10), 54–61.

Tiwari, S. & Gupta, A. (2015), 'A systematic literature review of use case specifications research', *Information & Software Technology* **67**, 128–158.

Tomayko, J. (1996), Carnegie Mellon's Software Development Studio: A five year retrospective, *in* 'Proceedings of 9th Conference on Software Engineering Education', IEEE Computer Society Press, pp. 119–129.

Truex, D., Baskerville, R. & Klein, H. (1999), 'Growing systems in emergent organisations', *Communications of the ACM* **42**(8), 117–123.

Turner, M., Brereton, P. & Budgen, D. (2006), 'Service-enabled access control for distributed data', *IEE Proceedings – Software* **153**, 24–30. Special Section on RBAC.

Turner, M., Budgen, D. & Brereton, P. (2003), 'Turning software into a service', *IEEE Computer* **36**(10), 38–44.

Vale, T., Crnkovic, I., de Almeida, E. S., da Mota Silviera Neto, P. A., Cavalcanti, Y. a. C. & de Lemos Meira, S. R. (2016), 'Twenty-eight years of component-based software engineering', *Journal of Systems & Software* **111**, 128–148.

van Aken, J. E. (2004), 'Management research based on the paradigm of the design sciences: The quest for field-tested and grounded technological rules', *Journal of Management Studies* **41**(2), 219–246.

van der Hoek, A. & Petre, M. (2014), *Software Designers in Action: A Human-Centric Look at Design Work*, CRC Press.

van Vliet, H. & Tang, A. (2016), 'Decision making in software architecture', *Journal of Systems & Software* **17**(638-644).

Vessey, I. & Conger, S. (1994), 'Requirements specification: Learning object, process and data methodologies', *Communications of the ACM* **37**, 102–113.

Visser, W. & Hoc, J.-M. (1990), Expert software design strategies, *in* J.-M. Hoc, T. Green, R. Samurçay & D. Gilmore, eds, 'Psychology of Programming', Academic Press, pp. 235–249.

Ward, P. T. & Mellor, S. J. (1985), *Structured Development for Real-Time Systems*, Yourdon Press.

Weaver, P. L., Lambrou, N. & Walkley, M. (2002), *Practical Business Systems Development Using SSADM: A Complete Tutorial Guide*, 3rd edn, Pearson Education.

Weinberg, G. M. (1971), *The Psychology of Computer Programming*, Van Nostrand Reinhold.

Weinberg, G. M. & Freedman, D. P. (1987), 'Reviews, walkthroughs and inspections', *IEEE Transactions on Software Engineering* **10**(1), 68–72.

Wendorff, P. (2001), Assessment of design patterns during software reengineering: Lessons learned from a large commercial project, *in* 'Proceedings of 5th European Conference on Software Maintenance and Reengineering (CSMR'01)', IEEE Computer Society Press, pp. 77–84.

Whittle, J., Hutchinson, J. & Rouncefield, M. (2014), 'The state of practice in model-driven engineering', *IEEE Software* **31**(3), 7985.

Wieringa, R. (1998), 'A survey of structured and object-oriented software specification methods and techniques', *ACM Computing Surveys* **30**(4), 459–527.

Williams, B. J. & Carver, J. C. (2010), 'Characterizing software architecture changes: A systematic review', *Information & Software Technology* **52**(1), 31–51.

Wirth, N. (1971), 'Program development by stepwise refinement', *Communications of the ACM* **14**(4), 221–227.

Wood, M., Daly, J., Miller, J. & Roper, M. (1999), 'Multi-method research: An empirical investigation of object-oriented technology', *Journal of Systems & Software* **48**, 13–26.

Woodcock, J., Larsen, P. G., Bicarregui, J. & Fitzgerald, J. (2009), 'Formal methods: Practice and experience', *ACM Computing Surveys* **41**(4), 1–36.

Wu, M.-W. & Lin, Y.-D. (2001), 'Open source software development: An overview', *IEEE Computer* **34**(6), 33–38.

Yakimovich, D., Bieman, J. M. & Basili, V. R. (1999), Software architecture classification for estimating the cost of COTS integration, *in* 'Proceedings of 21st International Conference on Software Engineering (ICSE'99)', IEEE Computer Society Press, pp. 296–302.

Yin, R. K. (2014), *Case Study Research: Design & Methods*, 5th edn, Sage Publications Ltd.

Yourdon, E. & Constantine, L. L. (1979), *Structured Design*, Prentice-Hall.

Zhang, C. & Budgen, D. (2012), 'What do we know about the effectiveness of software design patterns?', *IEEE Transactions on Software Engineering* **38**(5), 1213–1231.

Zhang, C. & Budgen, D. (2013), 'A survey of experienced user perceptions about design patterns', *Information & Software Technology* **55**(5), 822–835.

Zhang, C., Budgen, D. & Drummond, S. (2012), Using a follow-on survey to investigate why use of the Visitor, Singleton and Facade design patterns is controversial?, *in* 'Proceedings 6th International Symposium on Empirical Software Engineering and Measurement (ESEM)', ACM Press, pp. 79–88.

Index